The Pragmatics of Mathematics Education

Studies in Mathematics Education Series

Series Editor: Paul Ernest, University of Exeter, UK

The Philosophy of Mathematics Education
Paul Ernest

Understanding in Mathematics
Anna Sierpinska

Mathematics Education and Philosophy
Edited by Paul Ernest

Constructing Mathematical Knowledge
Edited by Paul Ernest

Investigating Mathematics Teaching
Barbara Jaworski

Radical Contructivism
Ernst von Glasersfeld

The Sociology of Mathematics Education
Paul Dowling

Counting Girls Out: Girls and Mathematics
Valerie Walkerdine

Writing Mathematically: The Discourse of Investigation
Candia Morgan

Rethinking the Mathematics Curriculum
Edited by Celia Hoyles, Candia Morgan and Geoffrey Woodhouse

International Comparisons in Mathematics Education
Edited by Gabriele Kaiser, Eduardo Luna and Ian Huntley

Mathematics Teacher Education: Critical International Perspectives
Edited by Barbara Jaworski, Terry Wood and Sandy Dawson

Learning Mathematics: From Hierarchies to Networks
Edited by Leone Burton

The Pragmatics of Mathematics Education: Vagueness in Mathematical
Discourse
Tim Rowland

Studies in Mathematics Education Series: 14

The Pragmatics of Mathematics Education:
Vagueness in Mathematical Discourse

Tim Rowland

Routledge
Taylor & Francis Group

LONDON AND NEW YORK

First published 2000 by Falmer Press

This edition published 2012 by Routledge
2 Park Square, Milton Park, Abingdon, Oxon, OX14 4RN
711 Third Avenue, New York, NY 10017

Routledge is an imprint of the Taylor & Francis Group, an informa business

© 2000 Tim Rowland

Typeset in Times by Graphicraft Limited, Hong Kong

British Library Cataloguing in Publication Data
A catalogue record for this book is available from the British Library

Library of Congress Cataloging in Publication Data
Rowland, Tim, 1945-
    The pragmatics of mathematics education : vagueness in
  mathematical discourse / Tim Rowland.
        p.   em.-(Studies in mathematics education series : 14)
  Includes bibliographical references and index.
    1. Mathematics-Study and teaching.   I. Title.   II. Series.
  QA11.R67   1999
  510'.71-dc21   99-26513
                                                              CIP

9-780-750-71012-1 (hbk)

Contents

List of Figures and Tables	viii
Series Editor's Preface	ix
Preface	xi
Acknowledgments	xiii

1 Preview and Methodology — 1

Two Principles — 2
Aims and Themes — 4
A Context in Language Research — 5
Mathematical Conversations — 6
Contrary Trends: Framing and Sample Size — 11
Ethnography and Interpretation — 12
Interpretation and Participant Observation — 16
Summary — 18

2 Generalization — 20

Generalization and Inductive Reasoning — 21
What is Inductive Reasoning? — 24
Induction and the Mind — 26
Mathematical Heuristic — 32
Truth and Conviction: the Theory of Numbers — 34
Explanation and Proof: Generic Examples — 38
Generic Examples and 'Real' Undergraduate Mathematics — 41
Recollection — 47
Summary — 48

3 Perspectives on Vagueness — 50

Viewpoint: Mathematics and Mathematics Education — 52
Interface: Mathematics and Language — 56
Viewpoint: Philosophy of Language — 61
Modality — 64
Pragmatics and Vagueness — 66
Summary — 68

4 Discourse and Interpretation — 71

Reference — 72
Pronouns and Reference: Power and Solidarity — 73

Contents

Some Approaches to Discourse   78
Overview: Approaches to Discourse   93
Summary   94

5   **Pointing with Pronouns**   96
    The Informants   96
    The Exclusive 'We'   97
    'It'   99
    On 'You' and Generalization   109
    Summary   113

6   **Hedges**   115
    Method   118
    'Make Ten': Frances and Ishka   120
    Hedge Types in Mathematics Talk   123
    Hedges: the Taxonomy Revisited   138
    Hedges and Politeness   140
    The Zone of Conjectural Neutrality   141
    Summary   142

7   **Estimation and Uncertainty**   145
    Estimation   145
    Counting   148
    Enquiry Focus   149
    Method   150
    Data   153
    Observations   156
    Interpretive Framework   157
    Interpretation of the Data   160
    Gender Differences   167
    Modal Auxiliaries   168
    Prosody   168
    Summary   169

8   **Pragmatics, Teaching and Learning**   171
    The Informal Research Group   171
    Case 1: Hazel   172
    Case 2: Ann   178
    Case 3: Judith   181
    Case 4: Rachel   186
    Case 5: Sue   190
    Case 6: The Public Lecture   194
    Case 7: Open University Video   195
    Case 8: Jonathan   198
    Summary   205

9 Summary and Review   207
    Pronouns   208
    Modality, Hedges and Indirect Speech Acts   209
    The Zone of Conjectural Neutrality   211
    Validation and Classroom Application   212
    Interpretation of Transcripts of Mathematics Talk   215
    Conclusion   216

Appendix 1   Transcript Conventions   218
Appendix 2   Index of Transcripts   219
References   221
Author Index   235
Subject Index   238

List of Figures and Tables

Figure 1   A taxonomy of hedges   61
Figure 2   Politeness strategies   87
Figure 3   Percentage of children in each age band giving Marked
           responses to each of three tasks   154
Table 1   Adjacency pairs: correlation between parts   92
Table 2   Frequencies and ranks of words in the 'Susie' corpus and
           other selected language corpora   100
Table 3   Markers across seven school years in four bands   153
Table 4   Age-related differences in the use of Markers   160
Table 5   Boys' and girls' responses   168

Series Editor's Preface

Mathematics education is established world-wide as a major area of study, with numerous dedicated journals and conferences serving ever-growing national and international communities of scholars. As it develops, research in mathematics education is becoming more theoretically orientated, with firmer foundations. Although originally rooted in mathematics and psychology, vigorous new perspectives are pervading it from disciplines and fields as diverse as philosophy, logic, sociology, anthropology, history, women's studies, cognitive science, linguistics, semiotics, hermeneutics, post-structuralism and post-modernism. These new research perspectives are providing fresh lenses through which teachers and researchers can view the theory and practice of mathematics teaching and learning.

The series Studies in Mathematics Education aims to encourage the development and dissemination of theoretical perspectives in mathematics education as well as their critical scrutiny. It is a series of research contributions to the field based on disciplined perspectives that link theory with practice. This series is founded on the philosophy that theory is the practitioner's most powerful tool in understanding and changing practice. Whether the practice concerns the teaching and learning of mathematics, teacher education, or educational research, the series offers new perspectives to help clarify issues, pose and solve problems and stimulate debate. It aims to have a major impact on the development of mathematics education as a field of study in the third millennium.

One of the major areas of growth in mathematics education research concerns the interactions between language, linguistics and mathematics. Although it has long been recognized that symbols and symbolization, and hence language, play a uniquely privileged role in mathematics and its learning and teaching, the systematic application of linguistics and discourse theory to the field has been slow to develop. I have speculated that this may be due to two factors. First of all, the dominance of mathematics education by traditional 'in the head' psychological theories has meant that mathematical talk and writing have been backgrounded while mathematical thought and cognitions have dominated the field. Secondly, thinking about mathematics has until recently been dominated by absolutist epistemologies which suggest that the role of language in mathematics is merely to describe the superhuman realm of mathematical reality. Thus the correct use of language in mathematics is often taken to be determined by immutable canons of logic. One outcome of this has been, as Tim Rowland says, neglect of the study of pragmatics and spoken mathematical discourse, because of its essential imprecision.

However, current developments have brought some of these assumptions into question. The scientific research paradigm, usually accompanied by absolutist epistemological assumptions, is losing its dominance in mathematics education research. Instead, interpretive research perspectives are gaining in ascendancy, and these more often rest on fallibilist epistemologies. There is a 'social turn' in the psychology of learning mathematics, and language, text, discourse and their relationship with the social context have now become central areas of investigation. Fully developed disciplinary perspectives and theories are being imported from linguistics and discourse theory for use in mathematics education. Initially, this arose from methodological necessity, as an increasing range of empirical research studies analysed transcripts of speech, discourse and texts. But now language in mathematics education is an area of study in its own right. The present work is situated in this emerging tradition. It opens up a new area of inquiry, the pragmatics of mathematics education.

Precision is the hallmark of mathematics and a central element in the 'ideology of mathematics'. Tim Rowland, however, comes to the startling conclusion that vagueness plays an essential role in mathematics talk. He shows that vagueness is not a disabling feature that detracts from precision in spoken mathematics, but is a subtle and versatile device which speakers deploy to make mathematical assertions with as much precision, accuracy and confidence as they judge the content and context warrant. Paradoxically, he shows that without vagueness classrooms would reflect less accurately the methods and practices of research mathematicians. His study is located within a fallibilistic view of mathematics, and it suggests ways in which this can be effectively extended to its teaching and learning. All in all, this volume exemplifies perfectly the series philosophy of linking theory with practice, with the result that both are enriched. No other book in mathematics education deals with the pragmatics of discourse analysis and the issue of vagueness, and this authoritative inquiry is set to become the standard reference in the area.

Paul Ernest
University of Exeter
1999

Preface

This book brings together a set of personal concerns and passions related to mathematics, linguistics and education. The various intersections of these domains are evident in different sections of the text; at times, the discussion is deeply immersed in one of these three disciplines.

I began my career as a teacher for the wrong reason – not out of any deep concern for pupils as learners, but because it seemed to afford me the opportunity of being paid *to do mathematics* under reasonably agreeable conditions. Up to a point, this proved to be the case. I enjoyed an amazing (by current standards) degree of professional freedom of choice and action, even in a state comprehensive school. If the students did sufficiently well in the public examinations, nobody asked how they got there. In time, of course, it dawned on me that *learning* was a fascinating study in itself. This shift in my motivation for teaching began to crystallize after I began teaching undergraduate mathematics to prospective teachers in Cambridge. Some of these students are woven into the fabric of this book. I wanted them to 'think for themselves', to solve problems, to reflect on their strategies and insights, to experience the 'buzz' of – in some sense – making mathematics. I think they did, but I was struck by their vulnerability as I asked them to predict, to generalize, to explain. This kind of mathematics was by no means cold, neutral, but infused with risk and charged with feeling. It needed a different kind of classroom 'management'. I realized that attention to cognitive 'process' demanded an affective price. This is also apparent, in the book, in accounts of 'mathematical conversations' with primary (elementary) school pupils; these children provided the data that first led me to adopt my particular, pragmatic approach to teacher–learner discourse of a mathematical kind.

The mathematics that most fascinates me now can be identified as a product of human activity – that of 'learners' of mathematics, myself included. Mathematics can be encountered and created by anyone who is prepared to engage with experience, looking for regularity and in a spirit of curiosity. Indeed, Freudenthal (1978, p. 123) describes mathematical activity in terms of 'organising fields of experience'. This organizing activity includes abstracting and generalizing from particulars. The Cockcroft Report (DES, 1982) sympathetically records a complaint that 'mathematics lessons are very often not about anything' (para. 462). No pupil should be made to feel that mathematics is so divorced from life, yet it is necessary to point out that mathematics is powerful *precisely because* it is not about anything in particular, enabling diversity of application and encouraging speculation beyond the realm of personal experience. The child who takes the imaginative leap from data to conjecture is truly engaged in an act of mathematical invention. Two theoretical perspectives on this observation feed my conviction of its truth.

The first is Lakatos's fallibilism (Lakatos, 1976), which asserts and emphasizes the fluid nature of human mathematical consensus (local or global) through a dynamic process of refinement of definitions, theorems and proofs. Thus the things we call 'mathematics' are the outcomes of human dialectic, a relative and subjective form of knowledge, perpetually open to revision. The second is radical constructivism (Glasersfeld, 1995), which recognizes that knowledge is *literally* what we make of it. That which we 'know' we comprehend in terms of what we have actively constructed from experience, and we know it in terms of that which we already knew. It is important to note that neither Lakatos nor Glasersfeld denies the existence of ultimate truth or objective ontological reality, although they would claim that useful notions of knowledge and understanding are possible without dependence on, or reference to, the existence of such a reality.

Convention and formalism may provide the passing illusion of a bedrock of reality, but offer little prospect of unlocking children's mathematical invention. Piaget's notion of equilibration captures the insight that learning, viewed as a form of active and dangerous creativity, is neither comfortable nor secure, although it may be immensely satisfying. The first, hesitant step towards new knowledge is conjectural, and the language of conjecture is not precise, but vague. Such a claim demands explanation and justification; that is the purpose of this book, in which I examine the circumstances in which vagueness arises in mathematics talk, and consider the practical purposes which speakers achieve by means of vague utterances in this context. The empirical database consists mainly of transcripts of mathematical conversations between adult interviewers and children. For the most part, the pupils involved in the study were aged between 9 and 12, although the age range in Chapter 8 is very much wider.

I draw on a number of approaches to discourse associated with 'pragmatics' – a field of linguistics – to analyse the motives and communicative effectiveness of speakers who deploy vagueness in mathematics talk. This alien pragmatic domain is somewhat novel for mathematics education, and I give a tour of the territory in Chapter 4 prior to application in subsequent chapters. I argue that vagueness can be viewed and presented, not as a disabling feature of language, but as a subtle and versatile device which speakers can and do deploy to make mathematical assertions with as much precision, accuracy or as much confidence as they judge is warranted by both the content and the circumstances of their utterances.

Acknowledgments

Over the years, I have had the good fortune to associate with colleagues who have inspired and supported me, as well as offering me good advice. It would be invidious to attempt to list them, and I name just one. Knowing and working with Bob Burn helped to make me a better mathematician and perhaps a better teacher. As for this book, I am indebted to many pupils and students whose words became my data, and to the teachers who themselves became the Informal Research group: especially Judith Addley, Kevin Gault, Hazel Matthews, Ann Neale, Sue Ray and Rachel Williams. Very special thanks are due to David Pimm and Margaret Deuchar, who encouraged and sustained me throughout this project. My thinking about links between mathematics, education and language is inseparable from my affectionate thinking about them. Thanks also to my friend Paul Ernest: for coming second in 1979, and for being so positive about this publication nearly two decades later. I am grateful to (and for) my sons, Mark and Simon, who each made significant, sensitive and expert contributions to this research and to the preparation of the book. Finally, I thank my wife, Judy, who tolerates and even loves the man who spends too much time at the computer: who is the expert proof reader and critic of almost everything I write, and to whom I dedicate this book.

I am grateful to the following for granting reproduction permission: Carlton Television Ltd and Watchmaker Productions Ltd for the quotation from the Clive James Show; the BBC for quotations from Casualty and Desert Island Discs; Ablex Publishing Corporation for Figure 1, taken from R. J. di Peitro (ed.), *Linguistics and the Professions*, Norwood, NJ, Ablex pp. 83–96; Cambridge University Press for Figure 2, from P. Brown and S. C. Levinson (1987) *Politeness: some universals in language usage*, Cambridge, Cambridge University Press; Rutledge Hill Press for tip number 448 from H. Jackson Brown (1991) *Life's Little Instruction Book*, Thorson's.

1   Preview and Methodology

It [APU practical testing] afforded an opportunity to hold a prolonged mathematical conversation with a child. My understanding of children's thought processes when solving problems has been considerably extended. (A teacher, quoted in Foxman et al., 1980, p. 73)

A mathematics teacher, Judith, has introduced her class of 14-year-olds to a mathematical investigation about line segments between points on a square-dot grid. Some way into the activity, she asks Allan, one of the pupils: 'Right. Can you make any predictions before you start?' Allan considers the question, and answers: 'The maximum will probably be, er, the least 'll probably be 'bout fifteen.' A first reading may suggest that there is nothing unusual about this interchange; we could, indeed, find many more like it. But step back; view the interchange through the lenses of an anthropologist. If, as it seems, Judith's intention is to request information, why does her question address Allan's *ability* to supply it? And why is his answer, ostensibly a mathematical utterance, so notably devoid of precision? What may we infer about Allan's attitude to his prediction from the manner in which he formulates it?

I set out on this research with a clear aim: to access and describe the mathematical frameworks and private constructions locked away in children's minds. My concern was to uncover what they 'knew' and how they structured that knowledge. I saw this as the most likely kind of outcome (in the spirit of what I shall call the 'linguistic principle') of the mathematical conversations in which I planned to engage them. In other words, I began with my attention focused on 'transactional' functions of language:

That function which language serves in the expression of content we describe as *transactional*, and that function involved in expressing social relations we will describe as *interactional*.

Whereas linguists, philosophers of language and psycholinguists have, in general, paid attention to the use of language for the transmission of 'factual propositional information', sociologists and psycholinguists have been particularly concerned with the use of language to negotiate role-relationships, peer-solidarity, the exchange of turns in a conversation, the saving of face of both speaker and hearer. (Brown and Yule, 1983, pp. 1–4)

I suggest that the interactional function of language is paramount in the exchange between Judith and Allan. Their interactional purposes are coded in the form of the language they use in the question and in the answer. The purpose of this book is to elucidate how that linguistic coding works, and to demonstrate the didactic value of knowledge of that code. Michael Halliday leaves us in no doubt as to its educational importance:

> If we consider the language of a child, there is good evidence to suggest that control over language in its interpersonal function is as crucial to educational success as is control over the expression of content, for it is through this function that the child learns to participate, as an individual, and to express and develop his own personality and his own uniqueness. (Halliday, 1976, pp. 197–198)

The two categories of language function are not exclusive, and both are of the utmost importance in talk about mathematics. My initial and continuing interest in transactional elements is evident in this book, in my implementation of variants of Piaget's clinical interview in many conversations with children. This research orientation is most strongly represented in Chapter 5.

Thereafter, the analytical focus shifts towards interactional components of mathematics talk. I shall explore how speakers in such conversations show their concern for a number of pragmatic[1] goals, principally those to do with the saving of 'face' (Goffman, 1967).

Two Principles

My approach to this research is guided by two fundamental, related principles, which I name 'linguistic' and 'deictic'.

Linguistic principle: language is a means of access to thought. One corollary of this principle is that talk with children has potential for insight into the structure of fragments of their mathematical thinking. The roots of the principle are in Freudian psychoanalysis and its provenance as a research method in education goes back at least to Piaget and famously bears fruit in mathematics education in Ginsburg (1977). The work of Douglas Farnham (1975) belongs to a strand of work with an explicitly linguistic foundation. Farnham drew on contemporary work of Barnes, Coulthard and others on patterns of classroom interaction to account for the child's development of mathematical understanding in terms of social sense-making. The linguistic principle, as I use it, makes no claim to 'transparency' – that, in some direct sense, one person's speech is a direct channel for the undistorted communication of their thought to others. Such a view is incompatible with my position regarding the construction of meaning. Subsequently, I supplemented the linguistic principle with:

Deictic principle: speakers use language for the explicit communication of thought, and as a code to express and point to concepts, meanings and attitudes. As I began systematically to tape-record and transcribe teaching sessions and one-to-one interviews with children, I experienced a growing awareness (which I describe first in Chapter 5) that the language the children used when talking about mathematics was of considerable interest in its own right – not in the sense that I had originally expected (for example, by providing descriptions of images), but in the subtle ways that the children used language to point to private concepts, meanings, beliefs, feelings or attitudes in the context of their mathematical thinking. I try to capture the essence of this language function in the term 'deictic', which derives from the Greek *deiknumi*, meaning 'to show' or 'to point'. The Greek root is associated with a linguistic term, 'deixis', which features in Chapter 5.

The centrality of the linguistic and deictic principles to my research orientation is examined later in this chapter, in a discussion of the clinical method, and in Chapter 4, where I shall set out some linguistic interpretive tools.

The deictic principle is at the heart of a paper in which the linguist Michael Stubbs (1986) draws attention to some ways in which speakers use language to convey beliefs and attitudes, or to distance themselves from the propositions they make. (A new presentation of the paper is available in Chapter 8 of Stubbs, 1996.) This epistemic subtext is sometimes summed up in the phrase 'propositional attitude' (Ginsburg et al., 1983, p. 26), glossed by Sperber and Wilson (1986a, pp. 10–11) as follows:

> Utterances are used not only to convey thoughts but to reveal the speaker's attitude to, or relation to, the thought expressed; in other words, they express 'propositional attitudes' [. . .]

Stubbs claims that no utterance is neutral with regard to the belief and commitment of the speaker, and urges the importance of the study of markers of propositional attitude:

> whenever speakers (or writers) say anything, they encode their point of view towards it: whether they think it a reasonable thing to say, or might be found to be obvious, questionable, tentative, provisional, controversial, contradictory, irrelevant, impolite, or whatever. [. . .] All sentences encode such a point of view [. . .] and the description of the markers of such points of view and their meanings should therefore be a central topic for linguistics. (1986, p. 1)

Stubbs identifies *vagueness* and *indirect* language as a principal means of encoding propositional attitude. The opening interaction between Judith and Allan exemplifies both indirectness (on Judith's part) and vagueness (Allan).

In fact, vagueness is the linguistic feature which unifies most of the data which I analyse in this book; rather, it is vague aspects of the language of participants in mathematical conversation that I shall single out for analytical attention. My principal reason for choosing that particular focus is that I came to see the significance, for mathematics talk, of Stubbs's insight about the encoding of propositional attitude. More surprisingly, I came to perceive how vagueness, suitably deployed, can also assist the transactional purposes of mathematics talk. The main and subordinate aims of my book are best understood in the light of these surprising perceptions.

Aims and Themes

My overall aim is to expose and understand some of the ways that participants in mathematics talk use language – especially vague language – to achieve their communicative and affective purposes. This comprehensive aim guides my choice of subject matter, and finds empirical expression in work that I present as four studies, reported in Chapters 5 to 8. Each study was motivated by a particular sub-aim, related to the main one.

First, in this chapter and in Chapter 4, I review the methodological and linguistic matters which underpin the design and interpretation of each of the four studies. Given the unifying theme of this book, I discuss some mathematical and philosophical dimensions of vagueness in Chapter 3. The mathematical process of generalization features strongly in three of the four studies; since I hold the view that this process encapsulates the essence of mathematical thought, I have devoted Chapter 2 to an exploration and exposition of its special character.

Chapter 5 is principally a detailed study of one 9-year-old child, Susie. My aim in that chapter is to demonstrate the transactional effectiveness of the pronouns 'it' and 'you' in our mathematical conversations. The vagueness of these words is associated with reference indeterminacy. I show how the first of these pronouns enables Susie to introduce certain concepts and generalizations into our conversation, despite the fact that she has no name for them. I demonstrate that the second is associated with vagueness-as-generality, and that 'you' surfaces in children's mathematics talk as a natural language pointer to generalization.

The study in Chapter 6 is based on several similar conversations with pairs of children aged 9 to 11. The similarity lies in the fact that each begins from the same numerical-combinatorial task, designed to provoke generalization. A paper of George Lakoff (1972) had first alerted me to a linguistic feature of the transcripts of these conversation, namely 'hedges' (such as 'I think', 'maybe', 'about' and 'around'). My aim in this study is to identify the use and prevalence of hedges in connection with conjectures. I show, as Stubbs suggests, that such hedges are powerful indicators of propositional attitude. In particular they point to vulnerability, they protect against loss of face. I shall introduce a construct which I call the Zone of Conjectural Neutrality, a space in which conjectures can be tested whilst minimizing the affective risk to their originators.

In Chapter 7, I report a study which aims to trace the development, between the ages of 4 and 11, of modal language competence, especially in the use of modal auxiliaries and hedges. The mathematical activities entailed in this study are counting and estimation. I identify a trend towards a developing ability to indicate propositional attitude in these ways in the primary years, and increasing awareness that vagueness is essential to estimation. I identify an anomaly in this trend, and account for it by reference to institutional factors.

The final empirical study, in Chapter 8, examines a disparate collection of teaching episodes across a wide age range, and involving eight different teachers. Here the aim is to validate three claims which arise from the findings in the earlier chapters. First, it is a demonstration of the applicability of the linguistic toolkit which I have assembled to the analysis of transactional and interactional features of transcripts of talk in mathematical interaction. Second, it confirms the prevalence and interactional significance of a number of previously identified (in the earlier studies) aspects of vague and indirect language in mathematics talk, across a wide age range. Third, by involving a group of teachers to work with me for that study, I was able to validate the relevance of my methods and findings to their day-to-day work in the classroom.

A Context in Language Research

The linguist Joanna Channell concludes her book on vague language (1994, p. 209) with a call for more research in 'variation study', including 'the study of occurrence of vagueness in different registers or genres'. The term 'register' refers to the specialized language peculiar to certain user-groups (Halliday et al., 1964). Such studies of aspects of vague spoken language exist in certain academic fields, for example medicine (Prince et al., 1982) and biology (Dubois, 1987). One aspect of variation not specifically included by Channell in her call for research is age-variation: the informants in every empirical study of vagueness of which I am aware are adults. This book makes a contribution to the study of vague language in these two dimensions of variation – register (mathematical) and age (especially 5- to 11-year-olds).

Mathematics education is necessarily and beneficially an eclectic discipline; the relevance of linguistics to this book is principally interpretive, in that certain organizing principles of language use, particularly those which have become associated with the young linguistic science of 'pragmatics' (Levinson, 1983; Mey, 1993), are applied to make sense of some vague features of mathematics talk identified in various corpora.

Hymes (1972) has claimed:

> Studying language in the classroom is not really 'applied' linguistics; it is really basic research. Progress in understanding language in the classroom is progress in linguistic theory. (p. xviii)

As it happens, most of the language data for my study were gathered in the course of 'interviews' in classrooms, rather than in naturalistic classroom settings *per se*. I see this study as basic research, not in linguistic theory, but in mathematics education. A number of regularities of speech found in the data will be described and interpreted as phenomena observable in the interaction between individuals as they talk about mathematics. Some evidence of the presence and broader consideration of the pedagogic significance of these phenomena in the interaction between practising teachers and pupils is provided in Chapter 8. The book concludes with some proposals for the application of this basic research in the cause of improving the teaching of mathematics, informed by my commitment to a constructivist view of learning and a quasi-empiricist philosophy, to be described in Chapter 3.

The ability of speakers to encode ideas, commitments, attitudes and beliefs in their utterances – not least in teachers' and pupils' mathematical utterances – would be of little use if their interlocutors were unable to decode, interpret and understand the intended subtext in such utterances. The communication and interpretation of propositional attitudes are central to mathematics education because the articulation of beliefs, conjectures and even 'answers' in mathematics is notoriously a risk-taking activity; this point is developed further in Chapter 6. Acknowledging that some relevant groundwork has been done in linguistics, David Pimm identifies interpretation, within classroom discourse, as an area which is now ripe for research effort:

> I predict the extremely subtle pragmatic interpretive judgements regularly made by both teachers and pupils in the course of mathematics teaching and learning in classrooms will move steadily to the fore as a research topic. (1994, p. 167)

In summary, this book could be viewed as

- a response to Stubbs's call for the description of markers of propositional attitude, in the specific context of mathematics;
- a contribution to variational study (Channell) of vague spoken language – in the domain 'mathematics'; and
- a partial fulfilment of Pimm's prediction concerning research in mathematics education.

Mathematical Conversations

For the most part, the data examined in this book consist of transcripts of mathematics talk. Much of this mathematics talk took place in the context of clinical interviews. In this chapter, I discuss the application of the interview method to research in mathematics education, and describe some principles which underpin my interpretation of the transcript data.

The 'linguistic principle' which I asserted earlier in this chapter reflects my confidence in the value of talking to pupils and students with a view to accessing their mathematical thinking. It is a conviction widely shared by researchers into children's thinking, deriving from a sense of the benefit of personal communication with the subject:

> It is my belief that the researcher can best formulate and test hypotheses and interpret the results of the tests in intensive interactive communication with the child, so that a close personal and trusting relationship can be formed. (Steffe, 1991, p. 178)

An approach to the study of children's thinking through 'interviews' with them is closely associated with Jean Piaget.

Piagetian Legacy

In 1920, at the age of 23, Piaget moved from postdoctoral study of philosophy of science and pathological psychology at the Sorbonne to a post at the Binet Laboratory in Paris (Piaget, 1952, p. 244).[2] His task was to standardize Cyril Burt's reasoning tests on Parisian children. In the administration of such tests, the wording and format of the questions were precisely defined, and had to be adhered to by the tester to safeguard the reliability of the procedure. Differences in performance between children (scaled by various 'measures') were calculated by reference to the correct responses given by them. Piaget, however, found the *incorrect* answers much more interesting, since they caused him to wonder what kind of reasoning gave rise to them. He realized, moreover, that such questions could not be researched by means of standardized tests.

> The first method that presents itself is that of tests [. . . in which] the question and the conditions in which it is submitted remain the same for each child [. . .] But for our particular purpose the test method has two important defects [. . .] the essential failure of the test method in the researches with which we are concerned is that it falsifies the natural mental inclination of the subject [. . .] The only way to avoid such difficulties is to vary the questions, to make counter-suggestions, in short, *to give up the idea of a fixed questionnaire.* (1929, pp. 3–4, my emphasis)

He concludes, having considered and dismissed both the 'test' method and 'pure observation', that a third approach may be deployed, one which exploits the best of each of the rejected methods whilst avoiding their disadvantages.

> This is the method of clinical examination, used by psychiatrists as a means of diagnosis [. . . in which] the good practitioner lets himself be led, though always in control, and takes account of the whole of the mental context. Since the clinical method has rendered such important service in a domain where formerly all was disorder and confusion, child psychology would make a great mistake to neglect it. (*ibid.*, pp. 7–8)

Piaget's interest in psychiatry originated in his mother's mental illness, which significantly coloured his own childhood (Piaget, 1952, p. 238), but in fact the (then) novel clinical methods of psychoanalysis were under wide discussion as to their educational application, and not only in Europe (Mackie, 1923).

Contingent Questioning

Piaget's introduction to *The Child's Conception of the World* (1929) contains his only discussion of the clinical interview as a research methodology. The method was subsequently developed and adapted (or 'revised' – see Ginsburg et al., 1983, p. 10) by Piaget from pure adult–child discourse to include manipulation of materials so that actions as well as words are added to the interpretive data bank. Piaget's classical work using the revised method is that on conservation, e.g. Piaget and Szeminska (1952). The clinical method became the basis of Piaget's work for half a century. Ginsburg (1981) and others (Ginsburg et al., 1983) argue strongly for the efficacy of the method in research into children's mathematical thinking

Ginsburg is perhaps best known in this field for his classic 1977 book *Children's Arithmetic*, in the preface to which he is explicit:

> The primary method is the in-depth interview with children as they are in the process of grappling with various sorts of problems [. . .] Interviews like these, involving close observation of individuals, are rare in mathematics education, but essential to improving it. (p. iv)

The clinical method is appropriate for the purposes of identifying (eliciting), describing and accounting for cognitive processes (Ginsburg et al., 1983, pp. 11–13). Such processes include prediction, generalization and explanation in mathematics. The description of 'intensive interviewing' in social science research, as given in Brenner (1985), has much in common with Ginsburg's account of the clinical method.

The characteristic dimensions of the verbal clinical interview[3] are:

- The interviewer employs a *task or tasks* to channel the subject's activity: typically, the interviewer presents a problem of some kind at the outset of the interview. Subsequent tasks or problems will depend on the subject's reaction to the initial task.
- The interviewer's questions are *contingent* on the child's responses: indeed, after the initial task has been presented, the interviewer's whole contribution to the interaction is judged and decided on the basis of the subject's contribution. That is not to say that the interviewer necessarily surrenders control of the interview (see my comments on 'frame' later in this chapter), but that s/he constantly makes instantaneous decisions about her/his questions and the direction of the interview.
- There may be some degree of *standardization*: the actual extent of standardization will depend on whether the interview is intended to discover or to elucidate cognitive phenomena. For example, concerning behaviour which has been previously identified and considered, standardization may assist 'explication' of behaviour or detailed study of the prevalence of some phenomena.
- The procedure demands *reflection*: the interviewer asks the subject to reflect on what s/he has done and to articulate her/his thoughts, typically by means of questions such as 'How did you do that?', 'Can you explain that?', and so on.
- The interviewer makes appropriate use of scientific *experimental method*, such as holding some variables constant whilst deliberately varying others. The contingent nature of the interviewer's responses enables her/him to test hypotheses that s/he has generated (either in the interview or as a

consequence of reflection on previous interviews with the same subject) to account for cognitive processes or other phenomena which have been identified in this interview, or in earlier interviews.
(based on Ginsburg et al., 1983, pp. 18–20)

Ginsburg insists that 'contingency of questioning' (1981, p. 6) is at the heart of the method; that the essential and distinguishing feature of the clinical interview mode is the contingent (responsive, interactive) nature of the interviewer's contribution.

The contingent interviewer is like a barrister in court, having continually to make rapid assessments of what 'witnesses' say, to probe without leading the witness. Unlike the advocate, of course, the role of the clinical interviewer is not supposed to be adversarial (winning a case) but analytical, striving to create the conditions for the surface manifestation – especially in speech – of the subject's thought.

The Use of the Clinical Method in Testing

A form of contingent questioning which is well known in the British context is the so-called Practical Testing mode used by the mathematics monitoring team of the Assessment of Performance Unit (APU) in the decade 1978–1987.[4] Alongside large-scale pencil-and-paper tests, the APU team developed a number of semi-structured, individual interviews based on practical tasks with weights, shapes, money, and so on. The purposes of these one-to-one interviews, which were unique among national assessment programmes, include 'exploration of children's reasoning and understanding of mathematical ideas' (Foxman et al., 1981, p. 4), an outcome which began to be stressed early in the APU testing programme. Despite the 'practical' label, the essential feature of this testing mode is that it is interactive. Having assigned a prepared task for the pupil, the tester (a teacher, trained for the role of 'practical tester') notes the child's responses. S/he may then offer 'prompts', to enable the pupil to reconsider an unprofitable strategy or to progress from a 'stuck' situation, and 'probes' to elicit or clarify the rationale underlying the pupil's response. For further discussion of the contingent questioning dimension of APU practical testing, see Rowland (1996).

'Long Practice' – Researching versus Teaching

The professional skills of teachers related to questioning ought to equip them particularly well to deploy this method, either for cognitive research or for diagnostic purposes. Indeed, the Department of Education and Science invested in the mid-1980s in the distribution of APU Practical Testing 'kits' to schools with this diagnostic purpose in mind. A clinical interview with a child may well result in learning *for* the child;[5] the child may even perceive such an interviewer as a kind of 'teacher'. The primary purpose of the interview is, however, *to inform the interviewer* about the child. The cultural obstacle for teachers is the improbable notion of a sustained mathematical discussion with a child which is not (by intention) in some way an improving experience for the child. As Lynn Joffe, a member of the APU Mathematics Monitoring Team, observed, there is a powerful temptation for teachers to teach:

> Although it is extremely difficult, and is asking a lot, testers are asked, as far as they possibly can, [. . .] to suspend their inclination to teach [. . .] One way of getting round the urge to teach, is to try and substitute non-directive questions where one might be tempted to teach. (Joffe, undated, p. 3)

Earlier, Piaget had noted a related, but different, problem for teachers:

> the clinical method can only be learned by long practice [. . .] It is so hard not to talk too much when questioning a child, especially for a pedagogue! (1929, pp. 8–9)

Piaget's remarks on the need for 'practice' convey the notion of the clinical interview (and in turn, by implication, the mathematical conversation) as a kind of art form, in which the artist (the interviewer) strives over time to develop and improve her/his performance. The analogy with questioning as a style of teaching is clear: an improvised, unique, oral 'performance' for which there are guidelines but no script. The development of the artistry through the study of tapes and transcripts is part of the satisfaction, for teaching as well as for researching. In the final chapter of her remarkable book *Wally's Stories*, an American kindergarten schoolteacher, Vivian Gussin Paley, explains how, in her classroom,

> the tape recorder preserves everything. It has become for me an essential tool for capturing the sudden insight, the misunderstood concept, the puzzling juxtaposition of words and ideas. I began to tape years ago [. . .] and I was continually surprised by what I was missing in all discussions. I now maintain a running dialogue with each tape as I transcribe its contents [. . .] *The tape recorder trains the teacher not the child*, who never listens to the tapes and who is curious about the machine only the first time. (Paley, 1981, pp. 217–218, my emphasis)

In Paley's book, episodes which are explicitly mathematical are the exception rather than the rule. She demonstrates, however, that – even in the routines of daily classroom events – talk, tape and transcripts can be a powerful means of researching and refining practice and of coming to understand children's thinking.

Conversations can be preserved as data, for later scrutiny, in the form of videotapes, audiotapes, field notes or transcripts (electronic or hard copy). Each of these media has advantages and disadvantages. The videotape, for example, includes non-verbal data (such as gestures and actions on materials) and seems to facilitate subsequent group consideration of features such as critical moments in the discourse; the audiotape preserves speech features such as intonation, pauses, and voice tone; the transcript, a transformation of the primary record of the event, focuses attention on the spoken word, or coded speech features. The transcript as electronic text file is invaluable to the computational linguist with an interest in (say) the relative frequency of use of certain words or grammatical structures. Since I have chosen to focus on spoken language, I principally audiotaped my data, and transcribed the tapes using a word processor.

Contrary Trends: Framing and Sample Size

Contingency and standardization are contrasting and, inevitably, competing dimensions of the clinical method. Both are related to, but not in direct causal relationship with, the notion of control. Bernstein has offered a theoretical construct which he calls 'frame' to capture the essence of the control of knowledge in the teacher–pupil relationship.

> This frame refers to the degree of control teacher and pupil possess over the selection, organisation and pacing of the knowledge transmitted and received in the pedagogical relationship. (Bernstein, 1971a, p. 50)

Frame is a form of boundary, in a given context, between what is to be included and what is to be excluded.[6] In the context of teaching and learning, where framing is 'strong', the fence around that which is to be learned is (supposedly) sharp, well defined. Where framing is 'weak', the boundary is blurred, fuzzy. Thus, for example, 'investigational learning' in mathematics would seem to require a weakly framed pedagogical relationship, since, outside a core (possibly but improbably empty) of intended content learning outcomes, it is expected and hoped that pupil activity will result in the acquisition of other kinds of mathematical and strategic knowledge.

As I have already observed, research interviewing is not, by design, teaching, but I find it helpful to borrow the terms 'weakly framed' and 'strongly framed' to identify poles in a continuum of control exerted by the interviewer over the content and direction of the interview. The standard 'method of tests' (Piaget, 1929, p. 3), in which the interviewer's questions are scripted and the subject's responses possibly coded, must lie at or close to one pole (strongly framed), in that the interviewer retains total control over the agenda. (Strictly speaking, the test designer has control *in absentia*, whilst the interviewer totally lacks control, since s/he has no discretion to deviate from the script.)

The empirical account in this book begins with just two children (Susie and Simon) and a sequence of extended one-to-one contingent interviews with each child. In no way are these interviews standardized; in each interview, only the initial task or question was pre-planned, and only one of the tasks was presented to both children. The interviews were weakly framed in the sense that, beyond the initiation gambit, I had no pre-set agenda of my own for these interviews, no prepared schedule of questions or tasks, since the aim was the *discovery* of phenomena and related intellectual processes (Ginsburg, 1981, p. 5). It could be said, therefore, that Susie and Simon had a significant share in the determination and control of the agenda for these interviews. A major outcome (for me, as researcher) was the identification of particular linguistic pointers (surface phenomena) to generalization (private mathematical process). These pointers are the subject of Chapter 5.

The next empirical stage in the research (Chapter 6) was designed to study the prevalence of such linguistic phenomena in relation to generalization and associated mathematical processes. The sample size was increased to twenty children, who were interviewed in pairs to facilitate peer interaction. Given the sharper enquiry focus of the interviews, a standardized interview agenda was planned in the expectation that each conversation would proceed in 'phases' leading to conjectures and, in some cases, attempts at the explanation of 'rules'. Contingency remained an important factor of my role as interviewer, allowing in particular for differences in interpretation of the initial task. Nonetheless, in comparison with one-to-one interviews in the first stage, the framing was stronger. The class of linguistic pointers which were identified for study at this stage are called 'hedges' in the linguistics literature.

The final empirical stage – rather, the final empirical stage at which I exercised any control over the agenda for the interviews – was designed to test a hypothesis concerning children's use of hedges in the context of the mathematical process of numerical estimation. This study is reported in Chapter 7. The design at this stage required a much larger sample, in the event the whole population (230 children) of one primary school. Each interview needed to be relatively brief, typically five to ten minutes, with the questions standardized and focused on three prescribed tasks. A small measure of contingency was necessary, depending on each child's initial responses to each of the three tasks, with corresponding prompts requiring the child to reflect on her/his responses. Otherwise, little deviation from the tasks was permitted. Thus the framing of these interviews was relatively strong, but not as strong as in a standardized 'test' interview.

Two trends are therefore inherent in the design and administration of these clinical interviews. Whilst sample size increases from two to 230, the interviewer's control over the agenda and the data he (in this case) gathers – corresponding to Bernstein's notion of 'frame' – shifts from relatively weak to relatively strong. In other words, as the sample opens up, contingency closes down. A wish to display these two contrary trends and to preserve the chronological order of events between 1991 and 1995 has influenced my decision to present the three studies in Chapters 5, 6 and 7 in that order.

Ethnography and Interpretation

Having accounted for the use of the verbal form of the clinical method as my principal means of data collection, I now sketch a broader framework of ideas which underpin my methodological commitment throughout the book. At the heart of this commitment is a belief that human events have no absolute 'meaning',[7] but that it is possible that they be made meaning*ful* (connected with other agreed meanings) both individually and socially. That is to say, meaning is dependent on interpretation, which in turn is shaped by the world-view of the interpreter.

> I maintain that 'critical reasoning' is an oxymoron, because consistent critical thinking shows that we are always inside our own vocabularies and our own angle on the world. We should give up the idea that we can somehow jump right out of our own limitations and achieve absolute knowledge, while yet remaining our- selves. (Cupitt, 1994, p. 20)

This is the perspective of interpretivism, a philosophical position often contrasted with logical positivism. Of course, the maintenance of personal sanity and inter- personal communication requires that social groups with a common interest – teachers, for example – normally go about their business as though consensual, interpreted meanings were absolute. This is the nature of inter-subjective knowledge.

Research into education, and mathematics education in particular, is necessar- ily an anthropological endeavour, entailing the study of the behaviour of members of *homo sapiens* by members of the same species. The advantages and disadvant- ages of this peculiar state of affairs are evident. Quasi-scientific methods of re- search, deriving from experimental psychology, with arm's-length collection of measurements from questionnaires and the like, may assist the researcher in achiev- ing 'objectivity' through emotional detachment from the fellow creatures whom s/he is studying. Yet this in itself may be insufficient for the researcher to gain critical insight into the phenomenon that s/he has identified for study (Fischbein, 1990, p. 11).

Qualitative methods are characteristically descriptive, inductive, speculative, interpretivist; drawing on naturalistic data such as recordings made in working classrooms, case studies, extended but loosely structured interviews, and from par- ticipant observation. Such methods permit the researcher to exploit his or her mem- bership of and association with the species or sub-species (e.g. 'teacher') which s/he is studying; to get close to, to make contact with, the context of study, or even to participate in that context.

Participation and Detachment

The research reported in this book sprang from cognitive ambitions. I never lost touch with those ambitions, nor did I lose my desire for insight into the mathemat- ical mind in action. But within a year the focus of my work had shifted from the cognitive in the direction of the affective; and from preoccupation with the indi- vidual as 'lone scientist' towards an interest in the interaction between individuals when they talk about and do mathematics. I shall make reference to linguistic forms in order to understand pedagogic interactions. This is *interpretive* in that the effort is oriented towards meaning-making, the goal is 'knowledge about social action within a context' (Kilpatrick, 1988, p. 98). The action that I study is speech, focusing on aspects of language that can, for one reason or another, be classified as 'vague'; the context is people talking about mathematics.

Margaret Eisenhart (1988) gives a thorough survey of the interpretivist research perspective, and discusses some implications for research in mathematics education. Eisenhart's article derives from an ethnographic perspective, but nonetheless confirms the conviction with which I began this research: that I would gain insight by interaction with the pupils, students and (as it turned out) the teachers whom I was studying.

> The purpose of doing interpretivist research, then, is to provide information that will allow the investigator to 'make sense' of the world from the perspective of participants; that is, the researcher must learn how to behave appropriately in that world and how to make that world understandable to others, especially in the research community. (Eisenhart, 1988, p. 103)

I also recognized that the holistic goal of interpretivist research would best be served by both affirming and drawing on, rather than denying, my personal and professional identification with the enterprise which I was studying – the teaching and learning of mathematics. Such familiarity necessitates a determined effort of detachment by the researcher at certain points in the research,[8] otherwise s/he is unable to discern anything at all remarkable about the events s/he observes, even at a phenomenological level. Paul Atkinson (1981), discussing the study of classroom language, writes that:

> The very familiarity of mundane, ordinary social activity can be a great barrier to analysis. [. . .] One has to work rather hard to make the effort of will and imagination to render what is familiar *strange*. One has to approach the data as if one were an anthropologist, confronted with a new, alien and exotic culture, and hence suspend one's own commonsense, culturally given assumptions. This is what ethnomethodologists mean by the task of making everyday life 'anthropologically strange'. (p. 100)

A tape recording is a permanent, transformed record of an ephemeral event which had a sound (here, principally speech) component. Atkinson speaks of the making of transcripts of classroom talk as a 'discipline' (p. 99), in that it forces attention to details of the talk such as hesitation, interruption, false starts and incoherence. The listener's brain is inclined to 'tidy up' such details as a gratuitous, sense-making kindness when talk is experienced as a purely auditory event. If I was present (usually as a participant) at the recorded event, a memory of the original context and the event remains. In the transcript, the word is made manifest as a random-access witness to the event. The text is now a transformed object in the world, tangible and *accessible for study in its own right*. For me, this transformation of tape into text is a significant means of achieving detachment from the interactions in which I participated.

A Spiral Process

The interpretive challenge is to translate knowledge of the text into knowledge about the participants in the original context. Professional linguists have paid little attention to mathematical discourse, perhaps because mathematics does not, on the whole, present itself as much of a social phenomenon, as compared with, say, doctor–patient interviews. But because conversation about mathematics *is* nevertheless a social phenomenon, it would be expected to have some linguistic features of human interaction in common with some other social situations.

The approach I have taken is substantially inductive, drawing on a precept of grounded theory (Glaser and Strauss, 1967) which recognizes and legitimizes the use of data as the *source* of research questions and hypotheses, not just a means of seeking answers to *a priori* questions and testing *a priori* hypotheses. For me, the transcript data are the source of my observations about linguistic phenomena in the text of the conversations. I may not know the 'technical' linguistic name of a phenomenon when first I notice it, or even know that it has a name. My interpretation of how and why these speakers use that feature of language in mathematical conversation, as an aspect of their communicative competence, is subsequently *informed* by related literature in pragmatics, as well as by the usual diverse range of knowledge (mathematics, mathematics education, general education, psychology, philosophy, sociology, and so on) which comprises the broad research-cultural base of mathematics education.

In an absolute sense, of course, every hypothesis about the world arises from experience, and in that sense from some kind of data. This observation is at the heart of the discussion of generalization in Chapter 2. In the context of educational research, René Saran (1985) presents some remarks on grounded theory methodology which forge particular philosophical links between this chapter and the next two, making reference to:

> three methods which I use repeatedly in a spiral-like research process – abduction, deduction and induction. (p. 228)

The term 'abduction' is due to Peirce (1934, pp. 99 ff.) and refers to the process of hypothesis *formation*; the human mind 'invents' and proposes meaning (expressed as a hypothesis) as an imaginative leap from the data. The researcher then applies deduction on returning to the data *'with new eyes*, to [. . .] order the facts in a new way' (Saran, 1985, p. 229), whilst needing to be aware of the dangers of 'an enchanting love affair with the hypothesis'. For example, the creation of a classification of selected features of the data (abduction) could have the effect of relating previously unconnected elements of the data. Peirce reserves 'induction' to refer to hypothesis testing – making comparisons between the data, as perceived 'with new eyes', and the previously unordered data, or indeed with additional data which was not used for abduction.

Linguistic theory plays a part in the process of abduction in this book. Awareness of regularities and theories of 'ordinary' language caused me to surmise that similar regularities might be found in mathematics talk and be accounted for in similar ways, but with a specifically mathematical dimension. Where this is the case, spoken mathematical interaction is indeed viewed with new eyes. The detailed application of the interpretive process is discussed further throughout the empirical sections of the book, especially Chapters 5, 6 and 7.

Hans Freudenthal has urged the possibility of viewing methodology as arising from *a posteriori* reflection on research activity.

> I don't remember when it happened but I do remember, as though it were yester-day, the bewilderment that struck me when I first heard that the training of future educationalists includes a course on 'methodology'. This is at any rate the custom in our country but, judging from the literature in general, this brain-washing policy is an international feature. Please imagine a student of mathematics, of physics, of – let me be cautious, as I am not sure how far this list extends – impregnated, in any other way than implicitly, with the methodology of the science that he sets out to study; in any other way than by having him act out the methodology that he has to learn! In no way do I object to a methodology as such – I have even stimulated the cultivation of it, but it should be the result of *a posteriori* reflecting on one's methods, rather than an *a priori* doctrine that has been imposed on the learner.
>
> I readily admit that the principle of 'learn first, apply later' works in educa-tional methodology no better than it ever did in mathematics; that is, where it works it does so to the benefit to a small minority of learners only – the future specialists of methodology. Yet, fortunately, the intimidated majority can count on the precious assistance of this authoritative guild, the pure methodologists, whose strength consists in knowing all about research and nothing about education. (Freudenthal, 1991, pp. 150–151)

Whilst Freudenthal is harsh on 'pure methodologists', his suggestion that the re-searcher should experience an intuitive sense of 'rightness' in her/his methodology, as an outcome of implicit 'impregnation', has been a guiding principle at various times when I deliberated how best to proceed with both collection and interpreta-tion of data. The interplay between action and reflection has a place in the progress of knowledge and understanding of all kinds. Justification of methodology, in par-ticular, may be the outcome of reflection on research activity.

Interpretation and Participant Observation

In particular, an interpretivist perspective on the data caused me to revise my initial view of my influence on and contribution to the data I had collected. In a summary of some characteristics of qualitative research, Burgess (1985, p. 8) suggests that 'studies may be designed and redesigned':

> All the methods associated with qualitative research are characterised by their flexibility. As a consequence researchers can turn this to their advantage, as a rigid framework in which to operate is not required. Researchers can, therefore, formulate

and reformulate their work, may be less committed to perspectives which have been misconceptualised at the beginning of the project and may modify concepts as the collection and analysis of data proceeds. (*ibid.*)

Perhaps the shift of perspective that most complicated my interpretation of the rich transcript data that I was obtaining was my eventual acceptance that my contribution to the clinical interviews was itself an object of interest. Hence the kernel of my attention, which initially was on the nature of children's mathematical constructs, shifted towards interaction, language and affect. This was, at first, a shift which I made with some reluctance. It initially came about because, when I spoke to others about my work, my talk would be about the *children*, the subjects of my interviews, with reference to fragments of transcripts which I had distributed. Invariably, before long, someone would turn the discussion to some aspect of my contribution as interviewer or 'teacher'.[9] It seemed pointless to deny that my part in the conversation had some influence on the child's, yet I had begun determined to be neutral in these clinical interviews, a mere channel (as it were) for the outpouring of the child's mathematical thinking. If that had been my expectation, it was not confirmed by peer group feedback.

My recognition of my influence as teacher substitute, simultaneous with my role as clinical interviewer/researcher, was double-charged. On the negative side, I could no longer claim to be eliciting some kind of 'pure' cognitive data from these children. Yet Piaget had never suggested that one could, and acknowledged the influence of the interviewer on the subject, at the very least as a person to be 'satisfied' with answers – Piaget (1929, p. 16) speaks of children 'romancing', sometimes inventing plausible, supposedly introspective accounts in order to produce an answer to the interviewer's question. On the positive side, I no longer had to defend my clinical interview technique as a flawless, quasi-psychiatric performance, but was at liberty to consider it critically as a quasi-pedagogic transaction in so far as it succeeded in managing the children's interaction with me and (in the study reported in Chapter 6) with each other, enabling children to predict, generalize and explain and to articulate their beliefs. Furthermore, if I analysed some transcripts as if they were teacher–pupil interactions (and not just investigator–subject interviews), there would be a greater prospect of drawing conclusions that could be relevant to other teachers with regard to their pedagogic interactions with pupils.

I do not, however, believe that this shift of perspective is in conflict with the standard guidance to clinical interviewers to resist the urge to teach. The shift of perspective came in the analysis of the transcripts, not in the method which guided my conduct of the interviews. The point is not that I viewed myself as teacher in these interviews, but that the children may have.

In this book, I begin as 'pure' contingent questioner (Susie, Chapter 5) and end as 'pure' teacher (Jonathan, Chapter 8). The fixed point, for the purpose of analysis and interpretation of texts, is that I must be researcher throughout. What I cannot claim as fixed is my perspective on my role – the study had to be 'designed and redesigned' (Burgess, 1985, p. 8). But a changed perspective is a fresh insight;

in that sense at least I have acted out and celebrated Freudenthal's precarious but liberating proposal (1991, p. 150) that methodology may be created in and identified from action; that 'methodology [...] should be the result of *a posteriori* reflecting on one's methods'.

Summary

This book will encompass a number of themes and related aims, with the superordinate aim of revealing and analysing some of the ways that participants in mathematics talk use vague language to achieve their communicative and affective purposes. Many of the transcript data that I analyse are obtained from contingent clinical interviews with children, which may be described as mathematical conversations. I have reported a shift of perspective in my analysis of the transcripts, (though not in the method which guided my conduct of the interviews) as a consequence of acknowledging that my contribution to the conversations must be one factor to be taken account of in the analysis. Whereas I did not view myself as teacher in these interviews, it is possible that the children may have.

One tenet which I hold constant, is belief in the central place and function of generalization in mathematics. In consequence, this process is central to the design and analysis of much of the book. I therefore devote the next chapter to an examination of the ways that generalization gives rise to the greatest delight and satisfaction in mathematical activity, and also to the greatest uncertainty.

Notes

1   In this book, I use variants of the word 'pragmatic' in two different, but related, technical senses, as follows. *Pragmatism* (or Pragmaticism) is the name of a philosophical position due to the American polymath Charles Sanders Peirce (1839–1914). It is intended to be a 'practical' (as opposed to 'theoretical') kind of philosophy, which embeds rational discourse in life and conduct. The name 'pragmatism' derives from the Kantian term *pragmatisch*, expressing relation to some human purpose. The essence of pragmatism is that human rationality and purpose are inseparable. Peirce's ideas have evolved through William James, John Dewey and others, and are central to the philosophical foundations of the interpretivist research paradigm (Giarelli, 1988). *Pragmatics* is the name of a branch of linguistics which attempts to interpret the meaning of utterances by reference to the motives of speakers and to context of use. The distinction between syntax, semantics and pragmatics goes back ultimately to Peirce's theory of signs, or semiotics (Lyons, 1977, p. 114). Indeed, Morris described pragmatics as 'the relation of signs to interpreters' (1938, p. 6), although these Peircean origins are now more or less irrelevant (Lyons, 1977, p. 119). The domain of pragmatics can usefully be viewed as those aspects of meaning that cannot be dealt with by means of truth-conditional semantics (Gazdar, 1979, p. 2).The adjective corresponding to both the philosophical position and the branch of linguistics is 'pragmatic'. I am also obliged to use the same word in the everyday, non-technical sense of real-world realism, utilitarian, sometimes in contrast to 'ideal'. The two technical meanings will be recapitulated and expanded when the context requires.

2   Amazingly, among some 10 million words of scientific, epistemological and psychological prose (and, for good measure, a 'philosophical novel' written at the age of 20),
Piaget wrote very little autobiography, and only twenty pages (Piaget, 1952) are available in English.

3   In fact, Ginsburg is comparing the dimensions of three related 'protocol' procedures:
talking aloud (without interviewer intervention), verbal clinical interview and revised
clinical interview. He argues that clinical methods are well suited to the requirements of
cognitive research involving children, and my list focuses on the characteristics of verbal
clinical methods.

4   It is supremely ironic that the APU, which had the ability to monitor national 'standards'
of pupil performance under the National Curriculum, was wound up by the British government in 1988.

5   Given the striking resemblance between clinical interview and Socratic dialogue, learning through reflection and enhanced awareness might well be expected.

6   The term 'frame' is used differently in discourse analysis, to mean an assimilative cognitive
structure, something like Piaget's 'schema'. See Brown and Yule (1983, pp. 238–241).

7   I should point out that this is not a theological statement, quite the contrary. Whilst
human beings are capable of 'insight', any ultimate meaning of things is bound to be a
mystery to finite intelligences. In other words, I do subscribe to St Paul's belief that 'we
see through a glass, darkly' (1 Corinthians 13:12). This is not quite the same as saying
that there is no such thing as ontological reality, but for practical purposes my interpretivist
position is entirely consistent with a constructivist view of knowledge, as presented in
the next chapter.

8   I take it as read that detachment is necessary in the interpretation of phenomena in the
interest of scientific integrity. My point (and Atkinson's) here is that analytical detachment (i.e. in isolating salient components of data) must be consciously exercised by the
observer who is a 'member of the tribe' which is the object of study.

9   At first I dismissed this interest in my contribution to the transcribed conversations,
believing that study of children is the key to the improvement of mathematics education.
I haven't changed my mind about this.

2 Generalization

Analysis and natural philosophy owe their most important discoveries to this fruitful means, which is called induction. Newton was indebted to it for his theorems of the binomial and the principle of universal gravity. (Laplace, 1902, p. 176)

I have had my results for a long time, but I do not yet know how I am to arrive at them. (Gauss, quoted by Lakatos, 1976, p. 9)

Much of the empirical work reported in this book is set in contexts where students of various ages are carrying out and talking about mathematical tasks. The precise mathematical content of the tasks is of less importance, for my purposes, than the mathematical processes in which the students are engaged. In order to understand what students say in such circumstances, and why they say things the way they do, it is important to understand the nature of the processes themselves.

Yet, it can be dangerous and unhelpful to draw a sharp distinction between process and content in mathematics. Arguably it is the content – numbers, shapes, groups, topological spaces, and so on – that most clearly distinguishes mathematics as mathematics, that marks it out from other domains of knowledge, for example science or history – whilst in both these cases there are content overlaps with mathematics. Time, for example, is a concern for all three. On the other hand, without the processes there would be no mathematics, or at least mathematics would have no products, no propositional content, no truths (theorems) about the objects of study – numbers, groups, and so on. Bell et al. (1983, p. 206) describe the process dimension of mathematics in terms of:

the style and atmosphere of the activity in the mathematics classroom [. . .] whether [pupils] see mathematics as a field of enquiry, or a deductive system, or a set of methods to be learnt from the teacher.

It is the activity, as opposed to the product of the activity, that picks out process aspects of mathematics. Hilary Shuard (1986, p. 104) lists some aspects of mathematics which a group of primary school teachers deemed to be 'processes'. It is a long list, and begins (gratifyingly, from the point of view of subsequent themes in this book) with: classifying, generalizing, predicting and estimating. The list concludes with some processes related to personal qualities: co-operating, working independently, persevering. Shuard comments:

It is interesting to note that these processes are all couched in terms of 'doing'; we do a process [. . .] processes are actions, or verbs. Some of these verbs, however, have related nouns which represent areas of mathematical content that express the same ideas. (p. 105)

A glib answer to the question 'What is mathematics?' is 'Mathematics is what mathematicians do'. Despite the inherent circularity of this statement, it is a definition that stresses the fact that mathematics is brought into being by human activity. As in any inventive endeavour, that activity survives only in the memory of the originator; the public legacy of the activity is invested in the product. One of my university mathematics teachers, Brian Griffiths, once put it like this:

> Over the course of time, some fabricated mathematics eventually becomes modified into a fairly permanent form. Its very permanence leads to its becoming highly prized, so that when people refer to 'mathematics', they have in mind only this content, or the fabricated mathematics, and they forget or discount the activity that led to it. (Griffiths, 1983, p. 298)

Processes which feature significantly in this book, especially in Chapters 5 to 8, are generalization, prediction, explanation and estimation. The discussion of generalization in this chapter comprehends aspects of prediction and explanation, and has unexpected links with estimation.

Generalization and Inductive Reasoning

Here are two tasks. Each proposes some activity – things to do – and poses a question or questions – things to think about in consequence of the activity.

- *Task 1. Partitions.* The integer 3 can be 'partitioned' into an ordered sum of (one or more) positive integers in the following four ways: 3, 2 + 1, 1 + 2, 1 + 1 + 1. Find all such ordered partitions of 4. In how many ways can other positive integers be partitioned?
- *Task 2. Reflections.* Draw two intersecting lines *l, m* in the plane. Choose a motif M (such as a capital F) and position it in the plane. Locate in turn the image M′ of M under reflection in *l*, and the image M″ of M′ under reflection in *m*. How is M″ related to M? Name a single plane transformation which maps M to M″. What happens if you choose other pairs of lines, other motifs and initial positions?

Suppose, then, that I carry out these tasks: I produce the activity, and consider the questions. Having done the tasks, I might ask myself the questions spontaneously, even had they not been explicitly stated – an act of curiosity as a consequence of the activity. The activities themselves yield 'data', secure but isolated 'knowledge', information-in-hand. The questions (spontaneous or explicit) cause me to look beyond the isolated items of information, to view that information as known samples from a (substantially) unknown class of phenomena, as specific instances of items in that class. The questions associated with the tasks suggest limitations to the class, although I may choose to work with restrictions or extensions of such limits.

In the case of the given Task 1, my information-in-hand will soon include the four given partitions of 3, the eight possible partitions of 4, and possibly the two partitions of 2. The question 'In how many ways can other positive integers be partitioned?' prompts the thought that my data belong to a class $P = \{(n, r(n)): n\varepsilon N\}$ where $r(n)$ is the number of partitions of the natural number n. My data consist only of the sub-set corresponding to $n = 2, 3$ and 4. What might the (infinite) remainder of the class be like? The information-in-hand is limited, and in any case is bound to be insufficient in itself for me to know the values of $r(n)$ for values of n beyond those for which I have data. It may, however, be sufficient to cause me to form some beliefs about those unknown $r(n)$ values, and I may be prepared to articulate such beliefs in terms of predictions or conjectures.

For example, I may observe that $r(n + 1) = 2r(n)$ for $n = 2, 3$ and hence predict that $r(5) = 2r(4)$, i.e. $r(5) = 8$ on grounds of cognitive systematization (Rescher, 1979). Such a prediction is amenable to confirmation. In this case, to achieve confirmation I would need to undertake identification and listing of the partitions of 5. Once confirmed, I might then go on to predict values of $r(6)$ and $r(7)$. I may anticipate the corresponding acts of confirmation with eagerness or with distaste as they become progressively (in this case, exponentially) more tedious. I may use a computer to automate the listing of partitions. It will all add to my sense of regularity in the system, but will always leave a countable infinity of unknown values of $r(n)$.

I may – indeed, I am likely to – go further than making a finite (in principle, confirmable) set of predictions, and make the following conjecture: that for all natural numbers n, $r(n + 1) = 2r(n)$. Since I know that $r(2) = 2$, I may formulate the conjecture as: for all natural numbers n, $r(n) = 2^n - 1$. These conjectures have the quality of generalizations; they are statements (of beliefs) about properties of an entire class, statements made despite the fact that the whole class has not been directly inspected and tested – indeed, could not be – for the property or properties in question.

Prediction can be viewed as a specialized form of generalization. Each feeds on the other, each is both parent and child of the other, although predictions are (generally) more straightforward to articulate, since they entail fewer quantifiers. As to the nature of mathematical generalization in terms of cognitive activity, an individual observes events, instances of some kind in a mathematical domain, of numbers or shapes perhaps. Some compelling desire to make sense in a holistic way, to integrate this set of information inputs – to impose regularity, to gain predictive power – seems to drive an involuntary unifying tendency, a generalizing force of the intelligence. One of the rewards of maturity can be delight in the awareness of insight. It is an intense physical sense of well-being, of things holding together, that I have committed to public writing only twice (Rowland, 1974, 1993). Anne Watson (personal communication, 1995) attempts here to say what generalizing feels like:

> I only know what it feels like to me when I generalise, I cannot say what it feels like to someone else, but I assume the mental action of generalising feels like something to other people. I found it very difficult to explain 'how to generalise' in words or actions when I was a classroom teacher, but quite effective to catch a moment when generalising seemed to be going on and suggest that children tried to hang on to the feeling of that moment, whatever it was, so that they might recognise it again in future. For me, I feel a different level of power, an approaching completeness, a different positioning of self when I generalise but I bet this won't be a universally useful description.

The kind of generalization I have described in the context of Task 1 is a classical 'pattern spotting' activity (Hewitt, 1992) – duly, meaninglessly and sometimes joylessly performed by the nation's adolescents since GCSE coursework institutionalized mathematical 'investigations' (Love, 1988, p. 250). The object of this kind of pattern spotting is to identify some mapping whose domain is the natural numbers.

Task 2, Reflections, is not of the same kind, for in that case a generalization associates a plane transformation with each pair of lines in the plane. The initial position of the motif affects its images, but the composite transformation, the 'product' of the two reflections, is independent of that position. That observation is itself a generalization; another generalization following from it might be a claim that (for intersecting lines) the composite of two reflections is always a rotation. At another level, the angle of the rotation can be related to the angle between the lines l and m. The associated class from which information-in-hand is available then consists of triples $(l, m, t(l, m))$ where $t(l, m)$ is the composite of reflections in l and m. The imaginative problem poser is aware of how the class can be extended; the judicious one knows about closing down parameters in order to highlight regularities. Generalization is a particular form of the epistemic phenomenon of induction,[2] and is properly and usefully considered from that broader perspective. The term 'induction' is derived from the Latin rendering (using *ducere*, to lead) of Aristotle's *epagoge* (epi-agoge, leading outside). The change of prefix, from out(side) to in(side), is interesting; inductive reasoning takes the thinker beyond (outside) the evidence, by somehow discovering (by generalization) some additional knowledge inside her/himself. The mechanism which enables an individual to arrive at plausible, if uncertain, belief about a whole population, an infinite set, from actual knowledge of a few items from the set, is mysterious. The nineteenth-century scientist William Whewell captures the wonder of it all:

> Induction moves upward, and deduction downwards, on the same stair [...] Deduction descends steadily and methodically, step by step: Induction mounts by a leap which is out of the reach of method. She bounds to the top of the stairs at once [...]. (1858, p. 114)

Here deduction is portrayed in terms of descent, just as the argument or syllogism is presented on the written page – methodical, steady, safe, sterile, descending. By contrast, induction is framed as daring, creative, ascending. Whewell discusses the complementary characters of induction and deduction (or 'demonstration'), and the symbiotic relationship between them. They must be 'processes of the same mind'.

Without induction there is nothing to justify by demonstration; but it is the business of deduction to 'establish the solidity of her companion's footing'.[3] We may describe the process of inductive reasoning and theorize about it, but it remains a mystery that an individual may confidently claim an infinite kind of knowledge from a finite kind of information base. The notion of 'knowledge of variability' (Holland et al., 1986) provides some insight into the conditions which seem to trigger and assist such inductive leaps into that which is unknown, yet which may be grasped with sufficient confidence to allow articulation.

What is Inductive Reasoning?

The philosopher Nicholas Rescher (1980) offers some ways of looking at inductive reasoning that have relevance to the data in Chapters 5 and 7. Rescher emphasizes that the crucial feature of induction is movement beyond the evidence in hand. It is a tool for use by finite intelligences, a solution to the problem of finding answers to questions on the basis of limited evidence. Questions requiring inductive solutions, such as 'What is the composite of any two given reflections?' are necessarily infinite in character. That is to say, the set of instances of the phenomenon which are the subject of the question must in principle be an infinite set to require an inductive solution. For if the set were finite, it would (in principle) be possible to compute every instance (member of the set) individually, thereby acquiring certain knowledge as to whether in every case such instances conformed to some supposed rule or regularity. Rescher (1980, pp. 8–9) goes on to analyse possible responses to the question: [Q] 'Is it the case that every F is also a G?' Mathematical examples include the following. Are all prime numbers odd? Is $n^2 + n + 41$ prime for all $n \varepsilon N$? Is the angle subtended at the circumference of a circle by a diameter always a right angle? Are all cyclic groups abelian?[4] Rescher leaves aside the motivation for asking such questions, i.e. the process by which they are generated as matters worthy of attention. The argument which follows is based on Rescher's epistemological analysis.

Suppose we observe a finite set of Fs. Suppose further that each observed F is indeed a G. On this evidence, we are bound then to agree that at least some Fs are Gs. Moreover, we have no evidence as yet that there exists any F which is not a G. We could just say 'I don't know' in response to Q. Whilst this is truthful, it is also evasive, adding nothing to either knowledge or belief. In the circumstances the response 'Yes, all of them are' intuitively presents itself as the best available, albeit provisional response; the optimal solution from the point of view of plausibility. I use 'plausible' here in the way that Polya (1992) uses the word – not in the sense of specious, but of pleasing, satisfying (Latin *plaudere*, to applaud, clap hands). Thus induction can be seen to represent a responsible form of cognitive 'gap filling', which supposes that our consistent sample of Fs faithfully represents the whole. This is not to claim that the solution 'Yes, all of them are' securely represents the truth, only that it qualifies as the best estimate to the truth which we are able to make on the basis of the evidence available.

An inductive inference can be viewed as an aspiring but failed deductive inference, in the following sense. (My choice of example is inspired by some enquiries of my former colleague, Rex Watson, into large gaps between consecutive prime numbers.) Suppose I set out to examine the claim (C, say) that no integer strictly between 1329 and 1360 is prime (Watson, 1994). This is equivalent to the conjunction of thirty individual propositions $\{P_i\}$ where Pi asserts that i is composite, and i takes every integer value between 1330 and 1359. From thirty premises $P_{1330}, P_{1331}, P_{1332} \ldots P_{1359}$ I am entitled to infer C, because it is a syntactic conclusion of those premises. Suppose I then proceed to test the truth of each member of the set $\{P_i\}$, for example by noting the divisor 2 for even values of i and then running divisibility checks for each of the ten possible odd prime divisors between 3 and 31 for each remaining odd value of i. I confirm that P_{1330} is true, P_{1331} is true, P_{1332} is true $\ldots P_{1359}$ is true. Since classical semantics is faithful to syntax, I am now assured of the truth of C. I have given a secure, deductive demonstration of it, in the form:

P_{1330}
P_{1331}
P_{1332}

$\ldots$

$\ldots$

P_{1359} (premises above the line)

C   (conclusion below the line)

In contrast, consider now the situation with inductive inference, typified by Task 1:

The number of partitions of 2 is a power of 2
The number of partitions of 3 is a power of 2
The number of partitions of 4 is a power of 2
The number of partitions of 5 is a power of 2
The number of partitions of 6 is a power of 2
<The number of partitions of any integer greater than 6 is a power of 2>

The number of partitions of every positive integer is a power of 2

The first five premises can be directly confirmed as true (by listing and counting), but do not by themselves justify the conclusion below the line. The plausible inference of the conclusion is enthymematic (information-extending); that is to say, the missing but necessary premise (shown inside the brackets < >) is tacitly supplied, thus presenting an inductive argument as if it were a deductive one. Since enthymematic premises are normally suppressed, there being no firm evidence on which to claim their truth, the induction has the appearance of a failed (incomplete) deductive inference. The missing premise is supplied in order to enable us to cross the 'epistemic gap' which separates the data from the 'answer' (to essentially infinitary questions). The epistemic gap is the 'residual distance' to be accomplished, requiring nothing less than an inductive leap. As I have already argued, the inductive conclusion represents the most satisfying solution, the most plausible truth-estimate available, and so provides a *post facto* justification of the undeclared addition of enthymematic premises.

This is not to say that the inductive conclusion has the status of certain knowledge. Nor is it simply an uninformed guess. It is a conjecture. As Polya says, discussing Goldbach's (unproved) conjecture that every even non-prime is a sum of two primes:

> We arrived so [from a finite data set] at a clearly formulated general statement, which, however, is merely a conjecture, merely tentative. That is, the statement is by no means proved, it cannot have any pretension to be true, it is merely an attempt to get at the truth. The conjecture has, however, some suggestive points of contact with experience, with 'the facts', with 'reality'. It is true for the particular even numbers 10, 20, 30, also for 6, 8, 12, 14, 16. (1954a, p. 5)

Polya's reference to an 'attempt to get at the truth' shares the same epistemic quality as my earlier reference to induction as truth-estimation in the erotetic enterprise. Observe that inductive inference is to be distinguished from stochastic (statistical or probabilistic) inference. Given that every F we have examined is also a G, I do not conclude (inductively) that most Fs are Gs. The inductive inference that every F (including the infinite set of F's which we have not inspected) is a G is an uncompromised, if provisional, commitment to regularity in 'nature'. The degree of commitment of an individual to the inductive conclusion, the extent to which they believe it to be true, may and does vary considerably. At the heart of this book is the study of how individuals are able to convey, by spoken language, the strength or fragility of inductive truth-estimates.

The notion of induction as truth-estimation forges the following association between generalization and estimation. Given a discrete, finite set, there exists a precise integer n_0 such that the statement 'there are n objects in the set' is a true statement when $n = n_0$; for all other values of n the statement is false. An estimate of n_0 is therefore, in common with an inductive inference, a pragmatic, optimal solution to an erotetic dilemma. In the case of estimation, an accepted, associated language of approximation exists, including words like 'around' and 'about'. This language is cultivated to express the awareness of the speaker that, whilst the claim that s/he makes is an estimate of the truth, in the classical bivalent sense it may actually be false (and is false if $n \neq n_0$). In the case of generalization, I shall show that speakers find their own pragmatic, linguistic means of conveying such awarenesess.

Induction and the Mind

A cognitive account of inductive reasoning, and generalization in particular, should include both a description of mental restructuring to include the acquisition of new knowledge (or a new way of looking at old knowledge) and an attempt to explain how the individual accomplishes the 'inductive leap' as a synthetic act. The literature seems to be strongest in the former aspect, and much of it relates in one way or another to the formation of concepts.

Abstraction

Skemp (1979) gives an account framed in broadly Piagetian terms. We possess cognitive schemas (networks of concepts) of various kinds which structure our perception of reality; in fact, they sensitize us to reality, but in a selective way. For example, I have a schema which includes the concept 'polygon' and links it with a number of spatial, numerical and aesthetic concepts. This schema has a selective influence on the way that I perceive spatial inputs, in that I am able to process them comfortably within the schema that includes 'polygon', and I will do so if at all possible. This schematic shaping of reality (or whatever we choose to call the incoming data) is 'assimilation'. But the concept 'polygon' came about (for me) by a process of 'abstraction' (*ibid.*, p. 24) so that I might include objects like pentagons and hexagons (less familiar) in a conceptual class along with triangles and quadrilaterals (more familiar). Note that the inclusion of certain objects entails the exclusion of others; concept formation requires certain qualities to be stressed whilst others are ignored. One consequence of this generalization (to the concept 'polygon' from a number of instances) is to bring within my reflective horizon an infinite class of concepts (including, for example, 22-gon and 469-gon) examples of which I have never seen, nor am ever likely to. Nevertheless I am able to state and prove theorems about such objects. Skemp calls such generalization (concept expansion) 'reflective extrapolation'. It has the quality of Piaget's notion of accommodation, which Skemp prefers to call 'expansion'. Thus, Skemp speaks of assimilation by a concept, and expansion of a concept.

It is a feature of inductive reasoning that the truth of an infinite (enthymematic) set of untested propositions is claimed, in order to expand and bind together a finite (usually small) set of items of data. The essential finiteness of the database of information-in-hand may be obscured by the manner in which it is obtained and presented. I am thinking here of the new generation of dynamic geometry software, typified by Cabri-géomètre. Suppose, for example, that I create a triangle and construct its three medians. I observe that the medians are concurrent. I vary the triangle by dragging one of its vertices on the screen. The concurrence of the medians is an invariant of every frame in the cinematographic presentation. The inductive generalization is readily made, and with conviction (Schumann and Green, 1994, pp. 85–86); perhaps because the software has enabled a vast set of confirming instances to be realized. Indeed, given the apparent continuity of the dragging process, the data set appears to be continuous, uncountable. This is an illusion, since the hardware design – pixels and the like – only permits a finite, though vast, set of configurations to be calculated and displayed.

Construction and Expansion

Harel and Tall (1991) observe that 'generalization' may refer both to a process (inductive thinking) and to the product (an inductive inference) of that process. They distinguish three different kinds of generalization:

- expansive generalization, the expansion of a schema without need for its reconstruction;
- reconstructive generalization, which occurs when a subject reconstructs an existing schema in order to widen its range of applicability;
- disjunctive generalization, the construction and addition of a new, disjoint schema to an existing one, to deal with a new context.

The growth of my own understanding of two ideas from group theory exemplifies these different kinds of generalization. Let H be a subset of some group G, and C1 the idea that I can calculate the cosets of H in G. Let C2 be the idea that it is possible to calculate the conjugacy classes of the group G. Having first encountered C1 (for the purpose of proving Lagrange's theorem on subgroups), I subsequently met C2 (perhaps in order to enumerate normal subgroups – I don't really remember). Initially I regarded C2 as a concept disjoint from C1. Both C1 and C2 were set in a schema of related examples and theorems – a case of disjunctive generalization. Somewhat later I took an interest in equivalence relations, was struck by the beauty of the Fundamental Theorem (that the equivalence classes induced by an equivalence relation form a partition), and became aware of the unifying significance of the set partition theorems associated with both C1 and C2. I could, moreover, write down equivalence relations R1 and R2 on the elements of G which induced the respective C1 and C2 partitions of G. In so far as I now perceived C1 and C2 as being examples of the same thing, this was expansive generalization. Many years later, in the study of transformations of real two- and three-dimensional vector spaces, I learned about group actions (in order to prove the orbit-stabilizer theorem). The partitions associated with C1 and C2 were both, I deduced, the sets of orbits of elements of G under suitably defined actions of G on itself. This reconstructive generalization, I felt, not only included C1 and C2 but made them somehow inevitable and special (in both senses of the word) cases. A group action will partition any set that it acts upon – including the group itself.

  Disjunctive 'generalization' barely merits the name at all, since it misses the opportunity for economy of intellectual effort and places a heavy load on memory. As Polya puts it:

> there are two kinds of generalisations, one is cheap and the other is valuable. It is easy to generalise by diluting; it is important to generalise by condensing. (1992, p. 11)

Harel and Tall argue that expansive and reconstructive generalizations are 'more appropriate for cognitive development', and that expansive generalization is the more straightforward of the two. I would comment that, in the context of inductive activity and learning, disjunctive generalization could relate to the accumulation of individual but isolated instances of a phenomenon; expansive generalization has some quality of conservative extrapolation, rather like a prediction of the next case on the basis of the previous cases; reconstructive generalization seems like inductive inference, so that each observed instance is viewed as a special case of a phenomenon with wide applicability. Harel and Tall conclude:

> In principle we believe that the most desirable approach to generalization is to provide experiences which lead to a meaningful understanding of the current situation, to allow the move to the more general case to occur by expansive generalization, but there are times when the situation demands a reconstruction and, in such cases, it is necessary to provide the learner with the conditions in which this reconstruction is more likely to take place. (1991, p. 39)

It may be fair to observe that mathematics education lacks a unified theory of generalization, although most mathematics educators will say they know it when they see it.

Intuition

How is it that humans (and other animals) manage to organize experience in such a way as to provide a basis for judgement about situations outside experience? The mysterious nature of this capability is nicely captured in the word 'intuition' – that which (at first, or never) we cannot rationalize, we label intuition. The Latin root of the word means 'to look inside', suggesting in-tuition, or teaching of/by the inner self. Fischbein (1987) uses 'intuition' to denote a 'type of cognition' by which we recognize some 'facts' about the world. Fischbein emphasizes a distinction between intuition and perception, the latter being awareness of objects and facts as a result of sensory inputs. Knowledge (more correctly, beliefs) gained as a result of both perceptions and intuitions may be false. For example, visual perceptions permit optical illusions; naive intuitions are the basis of laws of 'intuitive physics' such as 'velocity is proportional to applied force' (Orton, 1985; Champagne et al., 1980, p. 1077). It is worth adding that intuition can be an obstacle to the acceptance of truth; for example (Fischbein's), it is virtually impossible to accept intuitively that a set may be equivalent to one of its subsets (Dedekind's characterization of an infinite set). We exploit this – at least, I do – to play linguistic tricks on our students when we demonstrate that the mapping $n \rightarrow 2n$ defines an injection of the set $\mathbf{N}$ of natural numbers, and then ask them to agree that there are 'as many' even natural numbers as there are natural numbers. Tirosh (1991, p. 203) tested the plausibility of this particular infinite-cardinal 'paradox' and several others with students aged 11–17, finding 'incorrect' intuitions commonplace in the comparison of infinite sets. What is particularly interesting is her finding that student misconceptions were relatively stable across this age range. One of the great didactic challenges (not just with regard to intuitions about infinity) for mathematics education is the identification and confrontation of students' 'epistemological obstacles' (Bachelard, 1938; Cornu, 1991) – beliefs which become embedded in a knowledge schema because they function well in one domain of activity but which malfunction and lead to contradictions in another. Intuition can be viewed as a form of induction, in the sense that:

> intuition [. . .] always exceeds the given facts. An intuition is a theory, it implies an extrapolation beyond the directly accessible information [. . .] One may affirm then that intuitions refer to self-evident statements which exceed the observable facts. (Fischbein, 1987, pp. 13–14)

> It appears that intuition can be said to occur when an individual reaches a conclusion on the basis of less explicit information than is ordinarily required to reach that conclusion. (Westcott, 1968, p. 97)

In everyday use we perhaps stretch this to a point beyond induction, so that intuition, rather like premonition, advises us of facts and events in the absence not just of adequate evidence, but of any evidence whatsoever. Such intuitions may be ascribed to 'extra-sensory' forces and powers – in effect, extending the scope of the notion of sensory inputs, so that intuition becomes 'merely' a different kind of perception. By a criterion of inner confidence in the intuitive disclosure, however, this kind of cognitive experience seems not to be excluded by Fischbein:

> An intuition always exceeds the data on hand. However, being an extrapolative guess is not sufficient to define an intuition. A feeling of certainty is also a necessary characteristic of an intuition. Otherwise *it is a mere guess* [. . .] *The extrapolativity aspect is not always evident, because the apparent obviousness of intuitions hides the incompleteness of the information on which they are based.* (p. 51)

One could read the italicized (by Fischbein) comment in two ways. The 'extrapolativity' may not be evident in the (intended) sense that the individual may be unaware that the basis of evidence for their new 'knowledge' is incomplete, that it is indeed a generalization. Alternatively, in the case of 'mere' everyday intuition, and possibly in some cases of mathematical intuition, the extrapolativity may not be evident in the sense that the individual may be unaware that the new knowledge builds in some way on old experience; s/he may not consciously be aware of some deeply embedded kinds of evidence that s/he does in fact possess.

Constraints and Default Hierarchies

Peirce (1932) noted the ability of humans to exercise appropriate, pragmatic constraints in the kinds of questions they ask, so as to gain information which is not just new but (in some sense) worth knowing. He poses the question as to how individuals manage to ask the 'right' questions about the data (the fruit of experiences of every kind) available to them.

> Nature is a far vaster and less clearly arranged repertoire of facts than a census report; and if men had not come to it with special aptitudes for guessing right, it may well be doubted whether [. . .] their greatest mind would have attained the amount of knowledge which is actually possessed by the lowest idiot. (paras 2:752–3, pp. 474–476)

The matter of constraints (Holland et al., 1986) is also a factor in the calculation of how much data we need before a generalization can appropriately be made. Suppose I visit a remote Pacific island. I see a bird which my informant calls a 'shreeble', and it is blue. I am likely to suppose that (all) shreebles are blue. This is a conjecture, a provisional generalization. Suppose, however, that I see an inhabitant of the island whom my informant calls a 'Barrato', and he is obese. I am not likely to suppose that all Barratos are obese. Why the difference? As John Stuart Mill put it,

> Why is a single instance, in some cases, sufficient for a complete induction, while in others myriads of concurring instances, without a single exception known or presumed, go such a very little way towards establishing a universal proposition? (1843, p. 314)

One interesting answer to this question, proposed by Holland et al. (1986), lies in the notions of (1) default hierarchies of concepts, and (2) variability of objects with respect to others in the default hierarchy. In the example above, my default hierarchy recognizes the set of shreebles as a subcategory of the class of birds, and BLUE as a member of the category of colours. The observation of one blue shreeble activates the possibility that all shreebles are blue. I then calculate (i.e. make a judgement about) the extent of the variability of the superordinate category BIRD with respect to the superordinate property COLOUR, by reference to my knowledge of appropriate subordinates – robins, ravens, seagulls, parrots, and so on. In this case, my calculation suggests only modest variability in the BIRD–COLOUR relation, and so I attach some confidence to the conjecture that all shreebles are blue. A more sophisticated judgement might result from a more refined default hierarchy, e.g. choosing TROPICAL BIRD as the immediate superordinate category to SHREEBLE. The difficulty with the obese Barrato should now be evident: body shape (from skinny to obese) varies considerably among people of any given nationality – at least, there is sufficient variability for me to be unwilling to make any conjecture about Barratos from such a modest base of evidence. Indeed, it is unlikely that the observation of the single example (the obese Barrato) would even trigger the suspicion, the general conjecture.

Stamp (1984) recalls teaching a lesson on right-angled triangles. In the first two examples considered – (6, 8, 10) and (5, 12, 13) – it was observed that the area and perimeter had the same numerical value. This led to the conjecture that 'this happens every time'. Stamp reports that he 'denied' that it could be so, and in fact proceeds in the note to deductive demonstration that, with the exception of the given examples, the proposition is universally false! Why, given two confirming instances, was Stamp disinclined to formulate the conjecture? What prompted him to spontaneous denial that it could be true? The default hierarchy/variability analysis is certainly plausible here; mathematics teachers are very conscious of the confusion between perimeter and area, and very aware themselves that there is considerable variability between the two in fact. Had the class tried just one more right-angled triangle they would have been obliged to modify their hypothesis. One can speculate that anxiety that a false relationship may be inferred for all triangles or polygons explains Stamp's 'authoritarian' intervention.

What we believe or judge to be significant as a basis for generalization, in given circumstances, enables a distinction to be drawn between the logic and intuition of inductive confirmation. Consider the hypothesis 'All ravens are black'. The observation of a black raven is clearly a confirming instance, strengthening the conviction that the proposition 'All ravens are black' is true. But the hypothesis is logically equivalent to its contrapositive, which says that all non-black things are not ravens. Thus, on this logical account, it follows that a white shoe (for example) is a confirming instance, confirming and adding to the belief that all ravens are black. Yet, intuitively, it does no such thing. This dilemma is called 'Hempel's paradox' (Hempel, 1965). The approach to generalization through default hierarchies and variability accounts for, and justifies, the intuitive, sceptical reaction. For whereas (as with shreebles) we are in a position to judge the variability of the superordinates BIRD and COLOUR, the concepts NON-BLACK and NON-RAVEN have no meaningful superordinates, and we cannot even begin to make variability judgements.

Mathematical Heuristic

George Polya must take the credit for a revival of interest in mathematical heuristic in the two decades following the Second World War. Heuristic, by Polya's definition (1945, p. 102) is the study of the methods and rules of discovery and invention. Heuristic reasoning, he affirms, is in the service of discovery, and so is to be regarded as provisional and plausible; it is often based on induction, or on analogy. Polya codified heuristic methods and strategies; for example (1945, p. 103), if you cannot solve the proposed problem, can you imagine a more accessible related problem? Polya's *How to Solve it*, published in his fifty-eighth year, has all the freshness of a rediscovered art, and is still possibly the best known of his heuristic quintet (1945, 1954a, b, 1962, 1965). Unfortunately, it has been widely (mis)represented, in summary, as a four-stage recipe for problem solving:

- understanding the problem,
- devising a plan,
- carrying out the plan,
- looking back,

as if adherence to some heuristic algorithm were a sure path to a solution. The four-point list is too general to be helpful in the solution of actual mathematical problems; it is, I suggest, more realistically viewed as a generic framework for the *analysis* of problem solving rather than a prescriptive action plan. I concur with Burton (1984, p. 10) that problem solving cannot be taught, although one can over time acquire a few techniques (such as tabulating data) and a great many, less tangible, heuristic instincts (such as calling to mind familiar, analogous situations; see below). Perhaps Polya wants to correct the false impression that the processes of mathematical problem solving can be 'learned' when later, an old man, he writes:

> Solving problems is a practical art, like swimming, or skiing, or playing the piano: you can learn it only by imitation and practice. [. . .] if you wish to learn swimming you have to go into the water, and if you wish to become a problem solver you have to solve problems. (Polya, 1962, p. v)

> If he [the problem solver] possessed a perfect method, an infallible strategy of problem solving, he could determine the next step from the data of the incoming situation by clear reasoning, on the basis of precise rules. Unfortunately there is no universally perfect method of problem solving, there are no precise rules applicable to all situations, and in all probability there will never be such rules. (*ibid.*, p. 89)

It is clearer in *Induction and Analogy* (1954a) that what Polya is doing is encouraging the problem solver to recognize and reflect on their own heuristic, as well as on that of some distinguished mathematicians of the past, such as Pappus, Descartes, Leibnitz, Euler, Laplace, Bolzano.

> I tried to illustrate each important point [. . .] in several cases I was obliged to take a not too elementary example to support the point impressively enough. In fact, I felt that I should present also examples of historic interest, examples of real mathematical beauty . . . I tried [. . .] to give an appropriate opportunity to the reader for intelligent imitation and for doing things by himself. (Polya, 1954a, p. vii)

Polya goes on to stress the central place of induction as a paradigm for plausible reasoning, and continues:

> Observe also (what modern writers almost forgot, but some older writers, such as Euler and Laplace, clearly perceived) that the role of inductive inference in mathematical investigation is similar to its role in physical research. Then you may notice the possibility of obtaining some information about inductive reasoning by observing and comparing examples of plausible reasoning in mathematical matters. And so the door opens to investigating induction inductively. (p. viii)

Thus *Induction and Analogy in Mathematics* is a primer for the reflective, mathematical fieldwork that the reader is to undertake; it supplies 'the data for the inductive investigation of induction' which is to come in *Patterns of Plausible*

Inference (1954b). In the same way, *Mathematical Discovery* I (1962) is the reader's mathematical preparation for the cognitive exploration in Volume II (1965). This is the way of the inductive investigation of induction, and it is the method for the reflective investigation of mathematical problem solving. As William Whewell put it:

> For an Art of Discovery is not possible. At each step of the investigation are needed Invention, Sagacity, Genius – elements which no art can give. We may hope in vain, as [Francis] Bacon hoped, for an Organ which shall enable all men to construct Scientific Truths [. . .] this cannot be. The practical results of the Philosophy of Science must be rather classification and analysis of what has been done, than precept and method for future doing. (Whewell, 1858, p. v)

Truth and Conviction: the Theory of Numbers

The Theory of Numbers is notorious as fertile ground from which to generate inductive inferences; those (like Goldbach's conjecture) which deny as yet any counter-example, yet defy deductive proof, have a way of acquiring celebrity status. Perhaps Fermat's Last Theorem is currently the best known, given the interest in Andrew Wiles's epoch-making endeavours (Granville and Katz, 1993; Singh, 1997).

Polya (1954a, pp. 91–98) gives, *in extenso*, a translation of a memoir of Leonard Euler's. It is an account by Euler of his discovery of a 'formula' – a recursive scheme, in fact – to determine the sum $\sigma(n)$ of the divisors of (in principle) any positive integer n. Interest in $\sigma(n)$ derives, in part, from Greek fascination with 'perfect' numbers, such as 28 and 496, for which $\sigma(n) = 2n$. The story yields insight into the nature of belief and conviction, from the perspective of a mathematician with an unusual talent for invention.

Euler begins by remarking on the lack of orderliness in the sequence of prime numbers, and proceeds to an exposition of the well known method of evaluating $\sigma(n)$ from the prime factorization of n, using the multiplicative property of the function σ and the readily demonstrated fact that

$$\sigma(p^k) = \frac{p^{k+1} - 1}{p - 1}$$

for primes p. He then lists the calculated values of $\sigma(n)$ for $n = 1$ to 99:

1, 3, 4, 7, 6, 12, 8, 15, 13 . . . 120, 252, 98, 171, 156

Euler remarks that the list, like the sequence of primes, is a disorderly one (arguably more so, since it isn't even monotonic). Nonetheless he continues, 'I just happened to discover an extremely strange law,' to bring order to the apparent chaos. The 'law' is what we would call a recursive one, defining $\sigma(n)$ in terms of the set of preceding values $\sigma(m)$, for $m < n$. At first sight the scheme is as chaotic as the sequence it purports to explain:

$$\sigma(n) = \sigma(n-1) + \sigma(n-2) - \sigma(n-5) - \sigma(n-7) + \sigma(n-12) + \sigma(n-15) - \sigma(n-22)$$
$$- \sigma(n-26) + \sigma(n-35) + \sigma(n-40) - \ldots \text{ taking, if necessary, } \sigma(0) \text{ to be } n.$$

On inspection, the aspect of the scheme which is most obscure is the sequence 1, 2, 5, 7, 12, 15, 22, 26, 35, 40 . . . Euler explains that it will become clear by listing first differences: 1, **3**, 2, **5**, 3, **7**, 4, **9**, 5, **11**, 6, **13**, 7 . . . The patterns in the even and odd (placed) terms of this sequence is now evident, so that the list is indefinitely extendable.

It seems that Euler had derived his recursive scheme for $\sigma(n)$ from another of his number-theoretic conjectures, in the theory of partitions. In any case, he had no deductive proof that the law is universally true (i.e. for all $n\varepsilon N$). Euler freely admits this to be the case; the quotations which follow are from Polya (1954a, pp. 93–95):

> I must admit that I am not in a position to give it a rigorous demonstration . . .

It is now fascinating to examine what Euler considers sufficient grounds for belief in the generalization – for the reader and, presumably, for himself, since he has sufficient confidence to publish the result. He proceeds:

> it is not difficult to apply the formula to any given particular case, and so anybody can satisfy himself of its truth by as many examples as he may wish to develop.

Euler is saying that the finite set of confirming instances can be as large as the reader chooses, and anybody can satisfy 'himself' of their truth. He cannot, of course, establish (in a demonstrative sense) the truth of the formula (the generalization) by pointing to any number of examples, except by the addition of enthymematic premises. Euler, by reference to truth by, rather than of, examples, is claiming that, despite his admission that he cannot 'demonstrate' the truth of the formula, it will be possible to achieve conviction of its truth. Not content to leave it to the reader to generate some data, he says:

> I will justify it by a sufficiently large number of examples.

Clearly 'justify' must be about plausibility rather than proof. By modern standards of rigour, the great Euler stands guilty of naive empiricism (Balacheff, 1988, p. 218). What is 'sufficiently large' is a tricky question, but Euler implicitly suggests that 20 will do, and proceeds to recursive calculation, with his formula, of the values of $\sigma(n)$ for $n = 1$ to 20. In every case the value coincides with that given by direct calculation by means of prime decomposition. Was he just lucky? Certainly not:

> I think these examples are sufficient to discourage anyone from imagining that it is by mere chance that my rule is in agreement with the truth.

'In agreement' – the 'rule' and the 'truth' coincide for these 20 values of $\sigma(n)$. Yet awareness that, for example, $n^2 + n + 41$ is prime for $n = 1$ to 39, but not prime for all $n\varepsilon N$, can bring about a sceptical frame of mind, which Euler recognizes. In particular, the recursive calculation of $\sigma(20)$ only calls on six previous terms: $\sigma(n-1) + \sigma(n-2) - \sigma(n-5) - \sigma(n-7) + \sigma(n-12) + \sigma(n-15)$. He acknowledges the tentativeness behind his remark 'I think these examples are sufficient':

> Yet someone *could still doubt* whether the law of the numbers 1, 2, 5, 7, 12, 15 [. . .] is precisely that one which I have indicated [. . .] Thus the law could still appear insufficiently established and, therefore, *I will give some examples with larger numbers*. (Emphasis added)

He then picks 101 and 301 (the first prime, the second composite) and proceeds to confirm, as it were, agreement between the rule and the truth. For Euler, this constitutes a crucial experiment (Balacheff, 1988, p. 218). Surely, he seems to say, that clinches it:

> The examples I have just developed will undoubtedly dispel any qualms which we might have had about the truth of my formula.

Here Euler nicely exemplifies the role of *prediction* in relation to conjecturing activity. Its effect is to dip into the box of enthymematic premises, pick an item, and to examine either (1) whether it is in agreement with, i.e. an *extrapolative extension* of, the finite set of data-in-hand, or (2) whether it is a *confirming instance* of an already formulated generalization. The more random the choice, the more powerful is the epistemic effect of the confirming instance. It is like the celebrity drawing the winning raffle ticket who looks away from the box as she dips her hand into it, or the magician who rolls back his sleeves to show that nothing is concealed from the audience. The 'choice' of 101 and 301 is most interesting. With a 'live' audience, Euler could invite numbers to be tested; in written exposition he must appear to pick them randomly. In effect, he is saying, 'If it works for 301 it must work for anything.' The psychological thrust of this was illustrated for me by some first-year undergraduate mathematics students investigating the continued fractions of $\sqrt{n}$ for $n\varepsilon N$. Their findings were written up in project reports.

Emma notices that $\sqrt{3} = [1, \overline{1, 2}]$, $\sqrt{6} = [2, \overline{2, 4}]$, $\sqrt{11} = [3, \overline{3, 6}]$, $\sqrt{18} = [4, \overline{4, 8}]$. She makes a conjecture (inductive inference), tentatively expressed, and a conservative extrapolative prediction:

> Whenever n is of the form $r^2 + 2$ it seems as if the continued fraction is always of the form $[r, \overline{r, 2r}]$. Therefore I would predict that $\sqrt{27} = [5, \overline{5, 10}]$.

Emma proceeds to confirm her prediction.

Sarah generates data similar to Emma's, and makes the conjecture that $\sqrt{(n^2 + 1)} = [n, \overline{2n}]$ for all $n\varepsilon N$. Her prediction then proceeds:

I then picked a large value of *n* to check this general solution *to make sure it worked.* (Emphasis added)

She predicts that $\sqrt{82} = [9, \overline{18}]$ and proceeds to confirm it. For Sarah, the case $n = 9$ serves to confirm, to make sure of, all remaining instances. She suggests, as Euler might have done, that this is an example which will undoubtedly dispel any qualms that we might have had about the truth of her formula. Emma proceeds cautiously, her prediction being for the next value (5) of *n* outside her data. There is a strong sense here that she needs to reassure herself first, before she considers convincing others. I shall analyse my own experience of the need for such personal reassurance towards the end of this chapter.

Both students conformed to the normative requirements of proof, by algebraic transformation of their generalized surds, using the continued fraction algorithm. But Sarah, in common with Euler, offers us a 'crucial experiment' (Balacheff, 1988, p. 218) in preparation for (Euler – in place of) the deductive argument. The crucial experiment tests the plausibility of a generality by confirmation of an instance which is chosen for being not-special; a nice paradox.

I shall weave some personal remarks into a summary of the remainder of Euler's memoir. For me, the Theory of Numbers has been an inner laboratory, more like a playground, in which I experimented and theorized as an adolescent, read as a 'sixth form' student (when I first discovered the existence of books on primes, divisibilty and the like), studied as an undergraduate (personally tutored by Keith Hirst, since there was no taught course), and latterly returned to as a teacher. An undergraduate course that I taught for some years in Cambridge was inspired (in its content and the inductive approach to the content) by, and based on, a ground-breaking book by my former colleague Bob Burn (1982). The course includes a cul-de-sac topic on partitions (unordered, in distinction from Task 1). The sequence $\{p(n)\}$ of partitions of 1, 2, 3, 4 . . . is in fact 1, 2, 3, 5, 7, 11, 15, 22, 30 . . . The list has something of the impenetrable randomness of the sequence of primes. The course includes a lemma:

$E(n)$, the number of partitions of *n* into an even number of distinct parts, is equal to $O(n)$ the number of partitions of *n* into an odd number of distinct parts, unless *n* is of the form $\frac{1}{2}m(3m \pm 1)$; that is to say, unless $n = 1, 2, 5, 7, 12, 15, 22, 26, 35, 40 . . .$; in which case $E(n)$ and $O(n)$ differ by 1.

Whilst I had not encountered Euler's formula for $\sigma(n)$ before I read his memoir, I certainly recognized the sequence 1, 2, 5, 7, 12, 15, 22, 26, 35, 40 . . . , although I could not imagine such a direct connection between $\sigma(n)$ and $p(n)$. This situation exemplifies the unconscious activation of one element of Polya's heuristic catechism: 'Here is a problem related to yours and solved before. Can you use it?' (1945, p. 19).

In my case, the answer was 'No!'. It was by analogy (with the lemma on partitions) that I recognized the obscure pattern of Euler's sequence, yet I was unable to develop the analogy in order to achieve any explanation of Euler's formula for σ(*n*). As Pimm observes (1981, 1987, p. 100), analogy assumes a recognition of similarity between two situations, but has a *preferred direction* of application. My knowledge of p(*n*) offered potential for the illumination of σ(*n*), even if the potential was, for me, unrealized.

It turns out, however, that Euler had deployed proto-Polyan heuristic, and arrived at his 'extremely strange law' for σ(*n*) precisely because he had become aware of such a similarity. He did, after all, invent the theory of partitions. The memoir concludes with an ingenious but 'formal' proof of the 'law'. (Encumbered by our modern inhibitions about convergence, its validity would trouble us no end.) Assuming that the generating function[5] for E(*n*) – O(*n*) is $(1 - x)(1 - x^2)(1 - x^3)$ $(1 - x^4)\ldots$, Euler demonstrates that

$$-\left(\frac{x}{s}\right)\frac{dx}{ds}$$

is the generating function for σ(*n*), and thus forges the link between σ(*n*) and p(*n*). The remainder of Euler's argument is not so important for my present purposes, which is not exposition of the theory of partitions but the study of awareness of mathematical process.

Explanation and Proof: Generic Examples

In mathematics, proof may fulfil, at any time, one or more of a number of purposes. These purposes include not only assurance of truth, but explanation of (accounting for) observed regularities; and clarification of what it is that is being claimed (Hersh, 1993). For example, consider Task 1, Partitions. Seeking to account for the observation that the number of (ordered) partitions of each positive integer is twice that of the previous one, I argue as follows. Consider any partition of *n*. If I increase the size of the last part by 1, I have produced a partition of *n* + 1. If, instead, I adjoin an additional part of size 1, I have produced a second partition of *n* + 1. So there are at least twice as many partitions of *n* + 1 as there are of *n*. Finally, I make some remarks to assure that no partitions of *n* + 1 have been duplicated or unaccounted for. This constructive argument is very satisfying in that it explains why this remarkable and unexpected doubling phenomenon occurs. It is then easily adapted to a proof by Mathematical Induction that, for all *n*∈N, r(*n*) = 2^{n-1}.

Later, I have an entirely different insight. Imagine *n* as a sequence of *n* 1's, separated by *n* – 1 boundaries (one boundary between each consecutive pair of 1's). Each partition of *n* can be achieved by removing some of the boundaries, and 'gluing together' (adding) the 1's which are no longer separated by a boundary. For each of the *n* – 1 boundaries I have two options: remove it or leave it in place. Therefore there are 2^{n-1} partitions of *n*.

Polya (1954a, p. 114) remarks that this 'happens not infrequently', i.e. that a theorem which is proved first by Mathematical Induction is subsequently proved 'by some other method'. I would comment that my second proof is 'neat', economical, yet it leaves me little wiser about why I got the doubling pattern. Of course, I realize that 2^n is indeed double 2^{n-1}, but that observation seems to have no contact with the original problem about partitioning integers. This is at the heart of Hewitt's (1992) complaint that 'their [children's] attention is with the numbers and is thus taken away from the original situation'.

The first argument above (relating partitions of $n + 1$ back to those of n) is effectively presented, from the point of view of concreteness and conviction, by assigning a particular value to n, say 4. The exposition then describes how each partition of 4 begets two partitions of 5. Indeed, my experience with students indicates that careful scrutiny and comparison (with $n = 3$, say, for manageability) of the four partitions of 3 alongside the eight partitions of 4, can trigger explanatory insight concerning the way each partition of 3 is related to two partitions of 4. Such an argument amounts to proof by 'generic example' (Mason and Pimm, 1984; Balacheff, 1988).

> The generic proof, although given in terms of a particular number, nowhere relies on any specific properties of that number. (Mason and Pimm, 1984, p. 284)

The story (probably apocryphal, but see Polya, 1962, pp. 60–62 for one version) is told about the child C. F. Gauss, who astounded his village schoolmaster by his rapid calculation of the sum of the integers from 1 to 100. Whilst the other pupils performed laborious column addition, Gauss added 1 to 100, 2 to 99, 3 to 98, and so on, and finally computed fifty 101's with ease. The power of the story is that it offers the listener a means to add, say, the integers from 1 to 200. Gauss's method demonstrates, by generic example, that the sum of the first $2k$ positive integers is $k(2k + 1)$. Nobody who could follow Gauss's method in the case $k = 50$ could possibly doubt the general case. It is important to emphasize that it is not simply the *fact* that the proposition $1 + 2 + 3 + \ldots + 2k = k(2k + 1)$ has been verified as true in the case $k = 50$. It is the *manner* in which it is verified, the form of presentation of the confirmation.

Similarly, I could demonstrate, with a diagram, that $5^2 + (2 \times 5) + 1$ is a perfect square, in a manner that could convince that $n^2 + 2n + 1$ is a perfect square for *all* positive integers n.[6] As Balacheff (1988) so clearly and elegantly puts it:

> The generic example involves making explicit the reasons for the truth of an assertion by means of operations or transformations on an object that is not there in its own right, but as a characteristic representative of the class. (p. 219)

By way of contrast, consider the (false) proposition that $n^2 + n + 41$ is prime for all $n\varepsilon N$. I may confirm the truth of the instance when, for example, $n = 30$: by evaluating $30^2 + 30 + 41$, which is 971, and checking that no prime from 3 to 31 divides 971. But this gives no insight whatsoever as to why $n^2 + n + 41$ is prime for any other value of n. (The ambiguity of 'any' suits me well here; I want it to mean first 'some' (true) and then 'all' (false).)

Similarly, consider a possible inductive approach to Lagrange's Theorem[7] on groups. We might observe that the order of every subgroup of D_4 is a factor of 8, that the order of every subgroup of Z_{15} is a factor of 15, and so on. But these confirming instances lack *explanatory* power. The usual proof of the theorem enumerates cosets, and we could indeed demonstrate that (for example) Z_{15} is partitioned by the five cosets of the subgroup $\{0, 5, 10\}$ – but without gaining any insight as to why the cosets of subgroups of other finite groups should (a) be equinumerous and (b) partition the group. Any possibility of analogy with other groups would depend on structuring the presentation of the example in an appropriate manner.

Closely related, if not identical, to proof by generic example, is the notion of 'action proof' (Semadeni, 1984; Walther, 1984). The generic example serves not only to present a confirming instance of a proposition – which it certainly is – but to provide insight as to why the proposition holds true for that single instance. Walther indicates, for the validity of an action proof, the psychological necessity of the identification of aspects of special examples which are 'invariant regarding a transfer to other arbitrary examples'. (Walther, 1984, p. 10). The transparent presentation of the example is such that analogy with other other instances is readily achieved, and their truth is thereby made manifest. Ultimately the audience can conceive of no possible instance in which the analogy could not be achieved. In effect, the generic example triggers an inductive inference: that the argument holds in all cases. In saying this, I am suggesting that the generic example (suitably and skilfully presented) has the same role for proof as the confirming instance does for generating a conjecture. This may appear to be a fundamental methodological flaw in the method of proof by generic example. That is to say, it appears that the existence of a general proof is no more than a conjecture, an inductive inference in fact from the generic example. But this is to ignore the different demands of confirming instance and generic example, in the form of their presentation. The confirming instance has only to be demonstrated to be true, by any means whatever; the subtlety or lack of it in the demonstration is of no consequence. But in the generic example the demonstration must be more than a demonstration of truth; it must in some way explain, account for the property, in one instance, in the process of confirming it, so that that one instance is seen to be more than a case of serendipity.

Liz Bills (1996) attempts to capture this quality of generalization (from confirming examples) in the term 'structural generalization' (Bills, 1996; Bills and Rowland, 1999).

> A 'structural generalisation' generalises a result from a single or several examples *based on the generalisability of the process by which that result was obtained.*
> (Bills, 1996, p. 95, emphasis added)

In Bills's terms, structural generalization is the outcome of a cognitive act which sees the justification of the particular example as representative of a more general justification of a class of examples. This is in contrast to an 'empirical generalization' (Bills, 1996) of the kind lamented by Hewitt (1992) as mere pattern spotting without regard to cause.

We use the terms 'empirical' and 'structural' [. . .] we emphasise that one form of generalisation is achieved by considering the *form of results*, whilst the other is made by looking at the *underlying meanings*, structures or procedures. (Bills and Rowland, 1999)

This dependence of definition (of these terms) on the state of mind of the generalizing agent (person) in relation to the evidence and explanations available to him/her is undeniably precarious, but worth tolerating for the qualitative distinction made. Semadeni highlights a similar didactic dilemma concerning the acceptability of generic ['action'] proofs:

an action proof [. . .] involves a psychological question: how can one know whether the child is convinced of the validity of the proof by inner understanding and not just by being prompted by the authority of the teacher? Without dismissing this criticism, we note that it applies to any proof in a textbook: if the author finds his proof correct and complete, this does not automatically imply that students understand it. (Semadeni, 1984, p. 34)

The ability, indeed the tendency, of young children to explain by generic example is a feature of Chapter 6 of this book.

Generic Examples and 'Real' Undergraduate Mathematics

I have been interested, for some years now, in exploiting the generic example as a mode of explanation *that students are capable of generating* from their empirical experiences of mathematics, and also as a didactic tool *that I may choose to deploy* for some proofs that students would otherwise find convoluted in structure. I give an example of each of these kinds of use of generic examples. Both examples derive from my experience in Cambridge, teaching mathematics to well-qualified Mathematics/Education undergraduates. Their intention to become schoolteachers made it natural for me to introduce occasional discussion of their own learning.

'Stairs'

I taught a sixteen-hour course on mathematical processes for these students in their first term. The agenda for the course is much the same as that of John Mason's *Learning and Doing Mathematics* (1988). The problems which the students tackle vary from year to year, but I often introduced the well known 'stairs' investigation early in the course. The problem concerns the number of ways of ascending a flight of *n* stairs in combinations of ones and twos. The Fibonacci sequence readily emerges in the data, and the students are asked to consider why this is the case – in effect, whether the obvious inductive inference is as valid as it is persuasive.

Almost every time, one or more students eventually offered an account of why it is that the number of ways for any number of stairs will be equal to the sum of the number of ways for the previous two flights of stairs. The interesting feature of such student-generated accounts is that they are invariably grounded in a particular value of *n*.

But is such an argument perceived as generic by the students? On one occasion, seventeen students were present at a session when one of the group, Kim, volunteered an explanation as to why the number of ways for six stairs is equal to the sum of the number of ways for the previous two flights of stairs. I explained that I was interested in such arguments, issued a questionnaire (below), and asked them to complete it. There were spaces between the questions for the students to write their responses.

Climbing Stairs in Ones and Twos

Observation. The number of ways for six stairs (*thirteen*) is equal to the sum of the number of ways for five stairs (*eight*) and the number of ways for four stairs (*five*).

Kim's explanation. This is because, in ascending six stairs, the first step must be a one or a two. If it is a one, there are five stairs left, and we know that there are *eight* ways of climbing these five stairs. If it is a two, there are four stairs left, and there are five ways of climbing these four stairs. Therefore there are *eight* plus *five* ways of ascending six stairs.

1. Are you happy with the above explanation, i.e. is it convincing?

2. Does the above explanation help to convince you that the number of ways for fifteen stairs will be equal to (the number of ways for fourteen stairs) + (the number of ways for thirteen stairs)?

   If you answered YES, why does the explanation for six stairs convince you for fifteen stairs?

   If you answered NO, what would you need in order to be convinced?

3. Does the first explanation (for six stairs) convince you that the number of ways *for any number* of stairs will be equal to the sum of the number of ways for the previous two numbers of stairs?

If you answered YES, why does the explanation for six stairs convince you for any number of stairs? If you answered NO, what would you need in order to be convinced?

Of the seventeen students, the (anonymous) responses of fifteen indicated that the particular example – the explanation for six stairs – was indeed generic, for them, in relation to other numbers of stairs. The following responses are typical:

> *Student A, question 2.* If you start with a one, you have fourteen left. If you start with a two, you have thirteen left. So the sum of these two will form the same formula as for six stairs.

> *Student B, question 3.* We could rewrite the explanation in terms of n, $n - 1$ and $n - 2$, where $n = 6$ stairs in the explanation and $5 = n - 1$ and $4 = n - 2$. So we see we had the correct method in Kim's explanation.

Of the two who were unconvinced (questions 2 and 3), one was explicit about lack of conviction as regards the initial particular explanation (for six stairs). The other seemed to require further confirming instances of the generalization before s/he could 'accept it'.

Primitive Roots

The context is a 48-hour course which I taught in the Theory of Numbers, covering the usual undergraduate topics found in texts such as Baker (1984) or Davenport (1992). Building on the earlier 'processes' module mentioned above, I had begun to exploit generic examples as a forerunner of – sometimes instead of – conventional general algebraic arguments. Modular arithmetic arises early in the course, with the theorem that every prime number p has a 'primitive root' (that is to say, the group $\{1, 2, 3 \ldots p - 1\}$ under multiplication modulo p is cyclic). Unfortunately, confirming instances of this theorem give no clue as to how it may be proved. The standard general proof (see, for example, Baker, 1984, p. 23) is surprisingly indirect and overburdened with notational complexity. This latter factor is, in part, due to the fact that it has to deal with a double layer of generality: any prime p *and* any divisor d of $p - 1$.

For the first few times that I gave the course and 'taught' this general proof, I applied all my powers of exposition and explanation to make it accessible to students. I noted their difficulties and returned to my notes to hone and polish my presentation for the following year. Whilst my effort was well-intended, it was a case of self-delusion: deep down, I knew that I was adding to my own appreciation of the proof, but that the improvements were at best marginal from the students' point of view. Worse, the atomizing of the subtleties of the general argument into myriad tiny steps left students with no clear overview of the grandeur of the design.

One year, confronted with rows of puzzled faces at the completion of my exposition, desperation inspired me to add, 'Well, look, suppose p is 19, so that the order of M_p is 18. Now, what are the possible orders of the elements of M_p?' I continued, and as I did so it dawned on me that my choice of this *particular* example as a vehicle for the proof was enabling me to give a faithful account of the *general* argument. I soon began to theorize about the process, and then to call it proof by generic example. The next year, and the next, I *began* my exposition of the primitive root theorem with the generic example, and challenged the students to 'go home and write out the general proof'. Thereafter, I began to wonder why I was asking them to do that, since they would be able to do so only if their 'understanding' of the generality of the argument (borne by the generic example $p = 19$) was as complete as I could wish it to be.

Eventually, I carried out an enquiry similar to that on 'stairs' (reported above), with a class of second-year students on the same mathematics/education programme. The generic argument is summarized here, and omits many details which are considered in the exposition but which are not essential to my argument here concerning generic proof.

The prime number p = 19 has a primitive root

1. The group M_{19} has eighteen elements, so the order of each of those elements must divide 18. Possibilities are 1, 2, 3, 6, 9 and 18.
   Suppose there are N_1 elements of order 1, N_2 of order 2 ... N_{18} of order 18, so $N_1 + N_2 + N_3 + N_6 + N_9 + N_{18} = 18$. To prove that 19 has a primitive root, *we need to demonstrate that* $N_{18} \neq 0$.

2. Focus for the moment on the elements of order 6 (there may be none). Argue (as in the lecture) that if $N_6 \neq 0$, then $N_6 = \phi(6)$. A similar line of reasoning would establish that $N_1 = 0$ or $N_1 = \phi(1)$, $N_2 = 0$ or $N_2 = \phi(2) \dots$, $N_{18} = 0$ or $N_{18} = \phi(18)$. It follows that $[\phi(i) - N_i] \geq 0$ for each of the six possible values of i.

3. We know that
   $$\sum_{d|n} \phi(d) = n$$
   for all $n \varepsilon \mathbf{N}$, so
   $$\phi(1) + \phi(2) + \phi(3) + \phi(6) + \phi(9) + \phi(18) = 18.$$
   Since $N_1 + N_2 + N_3 + N_6 + N_9 + N_{18} = 18$, it follows that
   $$[\phi(1) - N_1] + [\phi(2) - N_2] + [\phi(3) - N_3] + [\phi(6) - N_6] +$$
   $$[\phi(9) - N_9] + [\phi(18) - N_{18}] = 0$$
   Since each bracket is non-negative, they must all be zero.

4. In other words, $\phi(1) = N_1$, $\phi(2) = N_2 \dots$ and in particular $\phi(18) = N_{18}$. Now $\phi(18) = 6$, so it follows that *19 has a primitive root* – six of them, in fact.

The students were asked to give a written response to three questions (analogous to those asked in the previous 'stairs' study):

1. Are you happy with the above explanation for the case $p = 19$, i.e. is it convincing?
2. Does the above explanation help to convince you that 29 has a primitive root?
3. Does the particular explanation [for the case $p = 19$] convince you that *every* prime has a primitive root?

As with the 'stairs' study, these students were asked to elaborate on their responses. The nineteen questionnaire returns indicate that the argument concerning the case $p = 19$ was generic for twelve of the students, whose responses included the following:

> [Concerning $p = 29$] The whole process could be repeated using 29 so that 1, 2, 4, 7, 14, 28 are possible orders, e.g. 7, same argument as for M_{19}. Come to same conclusions, you just have different numbers involved.

> I went through the proof with $p = 29$ and felt that it was applicable.

> You can adapt the proof so that it would apply to the possible orders of M_{29}, and then follow through the same argument.

> M_{29} has twenty-eight elements with divisors 1, 2, 4, 7, 14, 28 [. . .] $N_1 + N_2 + \ldots N_{28} = 28$, similarly $\phi(28) = 28$ because of the theorem [. . .] I can see that the argument can be transferred to $p = 29$ and would show that $N_{28} = \phi(28) = 12$. Therefore 29 has twelve primitive roots -- quite convincingly!

> [Concerning any prime] It is easy to follow the logical progression of the proof for $p = 19$ with any other prime in mind, and I can see no area of the proof which gives me any doubt that it wouldn't work for any prime.

Some respondents stated that they appreciated the 'concreteness' of the generic argument:

> If it was general, with no numbers, I think I would get confused.

Two students, moreover, volunteered that the generic argument with $p = 19$ indicated (for them) how they might formulate a general one:

> because you could just extend the argument for any prime and substitute in the values, and perhaps produce a general form for the proof.

> By changing 19 to p, I could generalize the argument.

For four of the students, the generic intention of the example was not effective. They were united by the sense that the proof for $p = 19$ was precisely that, and no more:

> Although the explanation for $p = 19$ is clear and true, it doesn't necessarily follow that $p = 29$ has a primitive root. So I'd prefer to work through it [$p = 29$] before I was convinced.

In effect, these students are being cautious about what they perceive as a case of empirical generalization (Bills, 1996). They fail to see that the argument has been presented with the aim of *structural* generalization; rather, the argument is not effective in suggesting structural generalization to them. It could be, of course, that these four second-year undergraduate students are conditioned into being satisfied with nothing less than a 'general' algebraic argument: indeed, they all indicated their need for a 'general proof' before they could be convinced. One student spoke for all four when s/he wrote:

> I would like a general proof to reassure me for all primes,

although s/he added somewhat wistfully:

> however, a general proof on its own would probably confuse me.

The ambivalence of the three remaining responses could not be resolved, since all the returns were anonymous; on the whole, these respondents found the case for primes other than 19 to be plausible but not entirely convincing. The distinction between public accountability and personal conviction is brought out in the following comment:

> It needs to be proved generally [. . .] but using a number 19 makes it easier to follow [. . .] A general proof needs to be followed through step by step. (But I am fairly convinced already – I can't see why it shouldn't work.)

As Walther (1984) indicates, the psychological effectiveness of the generic proof hinges on the identification and transfer of structural invariants, as opposed to contingent variables, in the argument. The sophisticated mathematician is able to isolate such invariants; indeed, some of the quotations from the students' writing above bring this out explicitly, such as 'You can *adapt* the proof,' 'The argument can be *transferred*.' One student uses the word 'similarly' to refer to this transfer from one particular case to another. The jury is still out on generic proofs in undergraduate mathematics.

I detect in the 'mathematical community' a commonly held view that generic proofs lack sophistication and are formally inadequate, at best a staging post between naive empiricism and general argument. I now question that conventional view, although I am aware that my position in relation to the possible sufficiency and completeness of generic proofs marks me out as an epistemological extremist. In the spirit of Semadeni (1984), I see no reason to give compliance with the demands of conventional symbolic formality priority over pedagogical responsibility to one's students. In my role as teacher, my responsibility is quite the reverse. I believe that learners of mathematics at all levels, including university students, should be assisted to perceive and value that which is generic in their particular insights, explanations and arguments. The barrier between such a level of knowing and the writing of 'proper' proofs is then seen for what it is – lack of fluency not with ideas but with notation.

Recollection

Time lends distance and detachment to one's ability to understand and integrate one's own past into the present. Before concluding this chapter I shall recount a personal but publicly documented event, and use it to illustrate and pull together some of the foregoing threads.

In Rowland (1974) I described a sequence of intellectual and domestic events leading to a mathematical insight (about regular polygons and generators of dihedral groups) which caused me great excitement. It centred on a sequence of polynomials and their roots. At one point I was stuck with the cubic $a^3 - 5a^2 + 6a - 1$. A solution and, indeed, a generalization, derived from the crucial insight that a solution of the previous polynomial, $a^2 + 3a + 1$, could be written as ψ^2, where ψ is the golden ratio, realized geometrically by the ratio [side of pentagram: side of pentagon]. I offered (p. 46) an 'elegant' geometrical confirmation of the truth of this insight before announcing the generalization that it suggested to me. I had no idea why the generalization should be true, I simply dared to hope that it could be – not (just) because I wanted to solve the problem that I had set myself, but because this solution had such beauty. I specialized the generalization I had made to three simpler cases, involving a regular hexagon, a square and an equilateral triangle, and was able to confirm them easily. If my generalization held, the square of the ratio [side of heptagram: side of heptagon] would satisfy the stubborn cubic. What I needed, at that moment, was merely to confirm that it did; for, if not, I had a counter-example to my generalization and the prospect of disappointment. The momentary prediction required confirmation without delay. In contrast to pentagons and hexagons, I knew nothing of the pure geometry of the heptagon; as I put it:

> Too excited to be sophisticated, I resort to trigonometry. [A diagram of a regular heptagon then indicates the angle θ between a side AB and a short diagonal AC.] $\theta = \pi/7$ so $(AC/BC)^2 = (2\cos \pi/7)^2 = 3.24$ approx., which [. . .] does seem to satisfy my cubic.

Aware of the blunt tools I had used for confirmation, I then asked:

> Can anyone demonstrate this more elegantly?

I was, perhaps, seeking a demonstration that had the quality of a generic example. The rest of the article is concerned with the plausibility of the generalization, but (at that time) I had no proof. I confessed:

> It has to be admitted that my 'conclusion' is rather a sweeping one, being pure conjecture based on six instances of the result. So far I have no proof [. . .] I need hardly affirm my belief in the conclusion, but as a piece of mathematics the work is incomplete . . .

The strength of my 'belief', in the total absence of proof, surprises me now. So strong was it that I was prepared to go into print, to put it on record. The editor,

David Fielker, did not hinder me. When one of my postgraduate education students, Alan Barnes, presented me with a proof just as *Mathematics Teaching* 69 went to press, I was pleased but not surprised that my belief was vindicated. In the end it was the beauty of the conjecture that assured me – nothing could, at the same time, have such beauty yet be false.

> Beauty is truth, truth beauty – that is all
> Ye know on earth, and all ye need to know.
>
> (Keats, 1820, 'Ode to a Grecian Urn')

Summary

For all learners of mathematics there is the possibility of acquiring new knowledge by reflection on appropriate and relevant experience (and arguably there is no other way). Generalization – unifying and information-extending insight – is central to such a means of coming-to-know, and may be viewed as a form of inductive reasoning. In the introduction to his inductive *Pathway into Number Theory*, Burn (1982) reminds us of an adage of Jacques Hadamard, that the purpose of rigour is to legitimate the conquests of the intuition. For the great mathematicians, as well as for novices, mathematics characteristically comes into being by inductive intuition, not by deduction. The products of induction are plausible truth-estimates. Therefore tentative belief, as opposed to certain knowledge, is an essential component of mathematical thought.

In the next chapter I shall link this observation with constructivist and quasi-empirical philosophies of mathematics learning, and begin to explore how vague language can be used to advantage in talk about beliefs and provisional knowledge about mathematics.

Notes

1 The first task is adapted from ATM (1967, p. 52), having also been adopted by the Assessment of Performance Unit (Foxman et al., 1982, pp. 102–111). It is an incredibly rich 'starter' for investigation, and I keep returning to it for work with students. The second task is more mainstream; see Fletcher, 1969, pp. 275–281.

2 Mathematicians are conditioned to associate 'induction' with Proof by Mathematical Induction, but I am speaking of induction here as a scientist would, in relation to discovery or invention. As I noted in Chapter 1, Peirce uses the word 'abduction' with similar meaning. Despite my appreciation of Peirce's maverick genius, I shall not adopt his term.

3 Whewell's personification of the characters Induction and Deduction (like the characters of Bunyan's *Pilgrim's Progress*) is delightful; it is as though they are two characters inhabiting the mind of the scientist. It is gratifying to note, moreover, that, Whewell, Master of Trinity College Cambridge, does not conform to the stereotype and make Induction female (illogical, intuitive, uncertain, apt to lead, to seduce her companion, capable of error) and Deduction male (logical, secure, the steadying influence on his partner). In Whewell's time it was taken for granted – certainly in Cambridge – that only

men 'did' science anyway, so it is all the more surprising that both characters are female ('She bounds to the top' . . . 'solidity of her companion's footing').

4 The formulation [Q] is a standard way of presenting inductive inference, and clearly well suited to questions such as 'Is every daffodil yellow?' In which case F is the set of daffodils and G the set of all yellow things. (A subset such as the set of yellow flowers also suffices.) If I inspect, say, ten daffodils (or even, with Wordsworth, a host of them), and find that each is indeed yellow, I am likely to reason inductively that every daffodil is yellow. Whilst [Q] seems to embrace a great many mathematical questions and propositions, it is not immediately clear that every mathematical question which might be answered by inductive methodology is necessarily of the form [Q], which is posed in terms of class inclusion. Nor is it clear that every conjecture which arises from inductive methodology is of the form [Q] (or, to be precise, a claim that every F is indeed a G). It seems particularly important to see whether the conjectural products of paradigm investigations such as Tasks 1 and 2 are accommodated by the Q formula. A possible solution is as follows. In the case of Task (investigation) 1, the conjectural outcome is a mapping g: $N \to N$; in fact, $g(n) = 2^n - 1$, but that is a detail. What is essential is the claim (conjecture):

[C]   for all $n \varepsilon N$, the number of partitions of n is in fact $g(n)$.

Now, let F be a set of integers, each of which is the number of partitions of some positive integer n. Formally, $F = \{r(n): n \varepsilon N\}$, where $r(n)$ is the number of partitions of n. Further, let G be the image of N under the mapping g, i.e. $G = g(N)$ or $\{g(n): n \varepsilon N\}$. The inductive conjecture C includes the claim that every F is a G, but C is much more specific; given any member x of F, C not only asserts that x is to be found among the elements of G, it actually specifies which element of G is to be identified with x. This, of course, is what mappings do, and it does appear to be a particularly demanding form of inductive reasoning, both in conception and in confirmation. Task 2, Reflections, can more easily (at one level) be cast in the mould of [Q], taking F to be the set of all composites of ordered pairs of plane reflections (in intersecting lines) and G to be the set of all rotations. Specifying precisely which element of G is to be identified with any given element of F requires the definition of a function t: $L \times L \to E$, where L is the set of lines in the plane and E is the Euclidean group of plane isometries.

5 The function $f(x)$ is said to be a generating function of the sequence $a(n)$ if, in the power series expansion of $f(x)$, the coefficient of x^n is $a(n)$. Euler 'knew' the generating function for $E(n) - O(n)$ only as an inductive conjecture, and makes much of this status in the memoir. He had inferred it himself inductively; it fell to others, later, to give a deductive demonstration, and to name it Euler's Theorem. A rare case, in Number Theory, of correct attribution. See, for example, Burn (1982, pp. 142–143). The lemma concerning $E(n)$, $O(n)$, combined with some formal manipulation of generating functions, leads on to Euler's Identity (Niven and Zuckerman, 1980, p. 274):

$$p(n) = p(n-1) + p(n-2) - p(n-5) - p(n-7) + p(n-12) + p(n-15) - p(n-22)$$
$$- p(n-26) + p(n-35) + p(n-40) - \ldots \text{ taking, if necessary, } p(0) \text{ to be } 1.$$

which now entirely resembles his 'law' for $\sigma(n)$.

6 An extensive and varied collection of such 'proofs without words' is assembled in Nelson (1993).

7 Lagrange's Theorem states that the order of (number of elements in) every subgroup of a finite group divides the order of the group. D_4 denotes the symmetry group of a square (order 8) and Z_{15} the set $\{0, 1, 2, \ldots 14\}$ under addition modulo 15.

3   Perspectives on Vagueness

> It is a mark of the educated man and a proof of his culture that in every subject he looks for only so much precision as its nature permits. (Aristotle, *Nicomachean Ethics*)

> Two plus two equals five – for sufficiently large values of two. (*Per* Eric Love, source unknown)

The first page of the 1982 *Report* of the Committee of Inquiry into the Teaching of Mathematics in Schools (the Cockcroft Report) included, in bold type, an assertion that:

> mathematics provides a means of communication which is powerful, concise and unambiguous (DES, 1982, p. 1)

and proposed the communicative power of mathematics as a 'principal reason' for teaching it. There was refreshing novelty in such a claim, which seemed to be justifying mathematics teaching in much the same way that one might justify the learning of a foreign language, and it did much to promote and sustain popular interest in the place of language in the teaching and learning of mathematics. Such a view of mathematics is in contrast, however, with that expressed in a more or less contemporary pamphlet issued by the Association of Teachers of Mathematics, whose authors argued that:

> Everyday speech is a highly tolerant medium. This tolerance is necessary because conversation is a form of action in the world; [...] Because it is a tolerant medium, everyday language is necessarily ambiguous.

> [...] Now, mathematising is also a form of action in the world. And its expressions, however carefully defined, have to retain a fundamental tolerance [...] Because it is a tolerant medium, mathematics is also necessarily an ambiguous one. (ATM, 1980, pp. 17–18)

This description of mathematics and conversation as forms of action emphasizes mathematics as human activity and suggests discourse as a means of communication, for mutual understanding and agreement. Furthermore, it offers the radical proposal that ambiguity is a beneficial ingredient in the formulation, the 'expression' of mathematics. As a *product* (polished, final), mathematics may be presented, particularly in writing but also in speech, as though it lacked ambiguity, representing truths about the world – or at the very least, about itself – in a sure, exact and unequivocal kind of way. This is tidy, but it is a deception of sorts. As Goguen puts it:

> Exact concepts are the sort envisaged in pure mathematics, whilst inexact concepts are rampant in everyday life. This distinction is complicated by the fact that whenever a human being interacts with mathematics, it becomes part of his ordinary experience, and is therefore subject to inexactness. (1969, p. 325)

The issue of vagueness has received relatively little attention in the literature of mathematics education, with the exception of the dimension of lexical ambiguity. The language with which mathematics is communicated and shared is rooted in natural language or 'Ordinary English', but which also includes elements of technical language – 'Mathematical English' (Kane et al., 1974) – much of which has Greek or Latin etymology and logical force (e.g. 'or' is interpreted inclusively). Rowland (1995a) analyses this blend, and some of the difficulties experienced by learners, in terms of mathematics talk taking place at two discourse levels – essentially object level and meta-level. The object level language bears the mathematical and logical substance which is capable of being coded in a formal first-order theory of mathematical logic. The meta-level language is not mathematics *per se*, but signifies the social interaction between people *doing* mathematics. It is what remains when the object level text is stripped away from the text of discourse.

Here, for example, 11-year-old Kerry and Runa are finding pairs of integers whose sum is 14. I have artificially identified the elements of the mathematical object language within brackets [*abc*]; they were in fact recorded on paper in symbols by the girls as they talked.

T8:73   *Runa:*   Let's put [fourteen], and then . . .
74   *Kerry:*   Right. [Ten add four.] Underneath.
75   *Runa:*   [Ten.] OK.
76   *Kerry:*   [Ten add four,] um [twelve add two]. Um [thirteen add one] . . .
77   *Runa:*   Wait a minute.
78   *Kerry:*   [Five. Nine add five.]
79   *Runa:*   Yeah, I was just thinking that.

The distinction between the two language levels, object level and meta-level, is often blurred because the mathematical object language incorporates and redefines many words from natural language. Ambiguities are inherent in words such as 'difference', 'similar' and 'or', whose mathematical meanings are related to their ordinary English meanings, but which cannot be inferred from them with adequate precision; they have to be learned within a process of mathematical enculturation. The clashes and ambiguities which result, especially for learners of mathematics, from this seemingly haphazard appropriation of ordinary English for mathematics is widely appreciated (Ullmann, 1962, chapter 7; Shuard and Rothery, 1984, chapter 4; Pimm, 1987, chapter 5; Barham, 1988; Durkin and Shire, 1991). One strand of the literature (Otterburn and Nicholson, 1976; Hardcastle and Orton, 1993) has highlighted the lexical deficiency of novices as users and and interpreters of mathematics at the object level, whereas I shall explore the interactive fluency of pupils in the use of mathematical meta-language.

Hans Freudenthal (1978, pp. 259–260) is critical of careless or spurious precision, and cites a Dutch encyclopaedia giving the length and weight of a lion as 2.40–3.30 m and 180–225 kg – data which betray their non-metric origin.[1] In another example, he points out that whereas, with regard to size, there is no significant difference between 10^{10} and $10^{10} - 1$, there is a world of arithmetical difference, in that one is divisible by 9 and the other is not. The difference between the two situations is *pragmatic*, i.e. related to purpose – the appropriate degree of precision depends on what you want to *do* with the number. Freudenthal points out the need for pupils to acquire judgement to distinguish between two worlds: 'the world where precision is a virtue, and the other where it is a vice, and [. . .] to be at home in both of them'.

These essentially pedagogic perspectives on mathematical and linguistic aspects of vagueness can be viewed as contributions to a debate concerning the place of precision, prescription and completion as factors in the acquisition of knowledge in general and mathematics in particular. In this chapter, I shall review some ways in which this debate is rooted in philosophies of mathematics and of language.

Viewpoint: Mathematics and Mathematics Education

The Cockcroft view of mathematics as precise and unambiguous reflects a popular view of mathematics, based on a tacit 'absolutist' philosophy. That is to say, the truths (theorems) of mathematics are sharp and certain, and in some way represent objective knowledge. Indeed, in this view, mathematics stands above and apart from empirical science in its purity and freedom from experimental error. Science can offer only 'theories', whereas the objects of mathematical thought and the assertions which are the products of mathematics are certain. Such a view is represented with passion and eloquence by G. H. Hardy in *A Mathematician's Apology:*

> mathematical objects are so much more than they seem. A chair or a star is not in the least like what it seems to be; the more we think of it, the fuzzier its outlines become in the haze of sensation which surrounds it; but '2' or '317' has nothing to do with sensation [. . .] 317 is a prime, not because we think so, or because our minds are shaped in one way rather than another, but *because it is so*, because mathematical reality is built that way. (1940, p. 130)

There is a comfortable sense of certainty in Hardy's words, but it reflects the state of mind of one who already knows. A 'fallibilistic' view of knowledge and truth offers a truer reflection of the experience of one who is coming to know.

Fallibilism

The austere perspective of absolutism, characterized by Hardy, contrasts with and is challenged by a fallibilist philosophy of mathematical knowledge. A latecomer to fallibilism, I have come to recognize it as an articulation of my pedagogical credo.

But first, it is reassuring to note that the rejection of absolutism is neither new nor irresponsible. Absolutism is linked, though not indissolubly, with the platonist's[2] belief that mathematical truth is 'out there', pre-existing and independent of human knowledge or lack of it. Each mathematical truth is, as it were, waiting to be discovered by the intelligence who, by genius or diligence, uncovers it. Over the last century, absolutism has been worked out in two major forms, logicism and formalism. The logicism of Russell and Frege attempted to reduce all mathematics to pure logic. Hilbert took the formalist view that mathematics is more than pure logic, but is capable of being axiomatized.

The arguments against absolutism (and, to some extent, against platonism) from within mathematical logic are essentially twofold. First, the deductive arguments which terminate in mathematical theorems must begin from a baseline of axioms, which are plausible products of observation or intuition. Any claim to absolute truth must then be suspect, since the very foundation is beyond the reach of demonstration. Second, truth begets truth according to an agreed (or tacit) set of logical axioms and rules of inference. Yet these rules are not beyond question or reproach, and alternatives to the classical scheme (first-order predicate calculus) include modal logic (Ackermann, 1956) and intuitionist/constructivist logic (Heyting, 1964).

The fallibilist critique of absolutism has been put forward in the writing of Imre Lakatos (1922–1973), notably in his posthumously published book *Proofs and Refutations* (1976). The book is explicitly set against the background of Polya's mathematical heuristic and Popper's critical philosophy of science (*ibid.*, p. xii). Central to Lakatos's critique is the failure of formalism to account for the growth of mathematical thought, either in peoples (phylogenesis) or in individuals (ontogenesis). Lakatos offers an alternative view of mathematics as the product of human mathematical activity and inter-personal dialogue.

> [. . .] informal, quasi-empirical mathematics does not grow through a monotonous increase in the number of indubitably established theorems, but through the incessant improvement of guesses by speculation and criticism, by the logic of proofs and refutations. (1976, p. 5)

The term 'quasi-empirical' mathematics refers to the observation that conjectures are the inductive outcome of consideration of 'data' collected in mathematical activity. An asymptotic refinement of definitions, theorems and proofs, argues Lakatos, is the outcome of human dialectic, acted out in the histories of cultures, and again (though not necessarily in the same way) in the classroom. In this 'fallibilist' view, mathematics is a relative and subjective form of knowledge, perpetually open to revision.

Paul Ernest's formulation and examination of fallibilism (begun in Ernest, 1991) was elaborated in a subsequent book (Ernest, 1998), in which the sheer originality of Lakatos's contribution is acknowledged – especially given that it goes against the stubborn grain of fixed views in a conservative discipline. In particular, the strengths of Lakatos's philosophy are identified as:

- clear and unequivocal rejection of absolutism in favour of fallibilism;
- reconceptualization of proof as an essential component of the context of mathematical discovery;
- a global (albeit partially developed) theoretical stance, offering insights for the history, philosophy, sociology, psychology and practice of mathematics;
- promotion of a social view of the development of mathematical knowledge.

> Lakatos's philosophy of mathematics [. . .] suggests a pattern for the development of mathematical concepts, conjectures, proofs and theories as a collective enterprise and that it indicated the role and variety of interactions contributing to this development. It shows that, contrary to some of the traditional accounts that abound, creation in mathematics is not essentially an isolated individual activity. (Ernest, 1988, p. 128)

The fallibilist position has been recognized and adopted by a number of other writers, whose stance is typified by these words of Reuben Hersh:

> It is reasonable to propose a new task for mathematics philosophy: not to seek indubitable truth but to give an account of mathematical knowledge as it really is – fallible, corrigible, tentative and evolving, as is every other kind of human knowledge. (1979, p. 43)

Whilst Hersh's comment is essentially epistemological, Sandy Dawson has explored the profound implications of Lakatos's quasi-empiricist philosophy for the teaching of mathematics. Writing about a 'fallibilistic way of teaching', Dawson has summarized his insight as follows:

> It was from ideas contained in Lakatos's articles and book that an alternative way of working in mathematics classrooms developed. [. . .] Lakatos claimed that the creation of mathematics comes about as the result of a process [. . .] in which a conjecture is created, tested and proved, or refuted and modified, or rejected outright. A classroom designed for pupils to operate in a fallibilistic fashion would provide pupils with a problem about which they could make conjectures as to its solution. [. . .] Opportunities to test and examine critically each conjecture must also be provided.
>
> A teacher who is functioning fallibilistically [. . .] establishes a classroom climate in which an atmosphere of guessing and testing prevails, where the guesses are subjected to severe testing *on a cognitive rather than an affective level* [. . .] where knowledge is treated as being provisional. Because of the provisional nature of knowledge, pupils are encouraged to confront the mathematics, their peer group and, where appropriate mathematically, even their teacher. (1991, p. 197, emphasis added)

Over the last decade John Mason has, in effect, been a consistent and effective champion of Dawson's pedagogic interpretation of Lakatos's fallibilist philosophy. Writing about the place of conjecturing in mathematical activity, Mason describes the qualities of what he calls a 'conjecturing atmosphere' in which 'every utterance is treated as a *modifiable conjecture!*' (1988, p. 9). Clearly, then, a fallibilist view *of mathematics* has implications for classroom conduct. This is certainly also true of a constructivist view of *learning*, which recognizes that knowledge is shaped by individual schemas and social frameworks of thought.

Constructivism

The precision and tidiness that characterize the public face of mathematics education, as it appears in mathematics textbooks for example, reflects the sense the author has made of it, or a sense negotiated and agreed by a group of people. This is not to say that meaning is arbitrary, or that any meaning is as 'good' as any other, but that individuals must construct meaning for themselves. A non-absolutist view of mathematical knowledge is implicit in recent formulations of radical (individual) and social constructivist epistemologies. Both assert the inevitability of the active sense-making role of the individual learner in assigning meaning to mathematical experiences (including the experience of being 'taught' or 'told') which, if not rejected altogether, are assimilated into his or her existing schema, or disturb and cause it to be revised. In this sense, no two individuals can 'know' one thing in quite the same way, although they might both assent to one linguistic or symbolic expression of it. The Piagetian roots of the radical epistemological theory are evident.

> Constructivism is a theory of knowledge with roots in philosophy, psychology and cybernetics. It asserts two main principles [. . .] (a) knowledge is not passively received but actively built up by the cognizing subject; (b) the function of cognition is adaptive and serves the organisation of the experiential world, not the discovery of ontological reality. (Glasersfeld, 1989, p. 162)

Glasersfeld (1995) has been at pains to emphasize that Piaget's epistemological position, and his own constructivist stance, derive not from a view about the means by which knowledge is acquired, but from a position concerning the status of knowledge itself. This is made explicit in the second of Glasersfeld's fundamental principles of constructivism, that 'cognition serves the subject's organisation of the experiential world, not the discovery of an objective ontological reality' (*ibid.*, p. 51). Essentially and literally, then, knowledge is what we make of it. Glasersfeld is very clear in his denial that knowledge is objective, absolute or pre-existent, and calls Piaget as his chief witness:

> Piaget's position can be summarily characterised by the statement: 'The mind organises the world by organising itself' (1937, p. 311). The cognitive organism shapes and coordinates its experience and, in doing so, transforms it into a structured world.

> What then remains is construction as such, and one sees no ground why it should
> be unreasonable to think that it is the ultimate nature of reality to be in continual
> construction instead of consisting of an accumulation of ready-made structures.
> (Piaget, 1970, pp. 57–58)

> Almost none of Piaget's writings could be fully understood without taking into
> account this revolutionary perspective. (Glasersfeld, 1995, p. 57)

Radical constructivism shares with intuitionistic constructivist philosophy a view of
the primacy of the individual in intellectual action and construction, and in its
apparent neglect (or lack of emphasis) of a cultural dimension to knowledge (Wilder,
1965, pp. 247–248). For a back-to-back discussion of the two 'constructivisms', see
Lerman (1989).

The social constructivist account sets individual construction of knowledge in
a context of convention, agreement and enculturation. Culture (both macro and
micro) is the broad *milieu* in which learning takes place and contributes to the
framework into which knowledge is integrated. It also emphasizes the role of language in the mediation and production of thought and in the development of
meaning (Bishop, 1985). In this social sense, it follows that 'negotiated' understand-
ings also possess a certain cultural relativity. There is an imperative in the social
account in the particular cultural context of institutional teaching and learning –
schools and universities – with affective as well as cognitive implications, such as
issues of status and 'face' (Johnson, 1970), dimensions which I consider later, in
Chapters 4 and 8.

Probing beneath its cultural and pedagogical manifestations, Paul Ernest has
undertaken the task of systematizing social constructivism as a philosophy of mathematics; for his wide-ranging and detailed discussion of the relationship between
objective and subjective knowledge of mathematics in social constructivism, see
Ernest (1991, 1998).

Interface: Mathematics and Language

A central concern for both mathematics and linguistics is the relationship between
form and meaning. It ought to follow that insights of each may potentially contribute to the understanding of the other. As an example of the interaction between
mathematics and linguistics, one with a formative and recurring influence in my
work, I now consider a linguistic application, to the vagueness-related problem of
'hedges', of the recent mathematical theory of 'fuzzy sets'.

Fuzzy Set Theory

Classical logic admits only two possible truth values for a statement, i.e. 'true'
or 'false'. A number of attempts have been made to extend the notion of truth in

order to accommodate vague, in-between states and concepts, such as drizzle and adolescence. Peirce had evidently entertained the idea of a three-value logic (Fisch and Turquette, 1966) and Lukasiewicz (1920) independently developed a triadic logic which attracted some attention.

A more radical solution to the problem of strict semantic interpretation of vague propositions involves a real-valued notion of truth, in which the value of a statement is a real number in the closed interval [0, 1]. The truth value of statement A is a measure of the *extent* to which A is true, with 1 and 0 corresponding to perfect truth and falsity.

An early attempt at such a solution was due to the philosopher Max Black (1937). Aspects of Black's approach have been developed more recently by Lotfi Zadeh, in the invention of fuzzy set theory (Zadeh, 1965). An electrical engineer, Zadeh was interested in the design of pragmatic solutions to complex problems such as pattern recognition or air conditioning – systems that 'worked' without requiring unrealistic or prohibitively expensive amounts of computing power. In other words, he was interested in ways of achieving solutions that are *good enough* as opposed to *exact*. He achieved this through fuzzy set theory, in which elements are deemed to belong to particular sets to a given degree.

Zadeh expounds his theory by reference to the fuzzy set (TALL) of tall men (*sic*). Suppose we agree that men less than 4 ft tall are members of the set TALL with degree 0, and that those over 7 ft are tall with degree 1. Zadeh then proposes a smooth curve (more an ogive than a straight line) joining the points (4, 0) and (7, 1). The degree of tallness of any person can now be read from the graph. Zadeh then extends this to set algebra; if T' is the complement of T and $m(T, x)$ is the extent of membership of x of set T, then $m(T', x)$ is defined to be $1 - m(T, x)$. Likewise $m(T \cup S, x)$ is defined to be $\max\{m(T, x), m(S, x)\}$. The valuations of all other Boolean functions of T, S, then follow from these definitions; an example is given later.

Zadeh's original paper was on fuzzy sets. Four years later, Joseph Goguen had worked out a corresponding fuzzy logic (1969), in which a statement A may be true, false or partially true, in so far as it is assigned a value ('degree') [A] in some partially ordered set $(L, <)$, the simplest example being the interval [0, 1].

For a survey of the theory and practice of fuzziness, see Kosko (1994).[3] Seven years after Zadeh's initial 1965 paper, George Lakoff published a linguistic application of the theory to hedges.

Hedges

Lakoff on Hedging

Some recent approaches to the problem of vagueness within the field of linguistics originate in consideration of the meaning and function of a class of words and phrases called 'hedges', which turn out to be central to my interpretation of vague aspects of mathematics talk in this book. Hedges include words such as 'sort of', 'about', 'approximately' – words which have the effect of blurring category boundaries or otherwise precise measures – as well as words and phrases such as 'I think', 'maybe', 'perhaps', which hedge the commitment of the speaker to that which s/he asserts.

The work of Zadeh (1965) and Goguen (1969) laid the foundations of fuzzy interpretation of vague language, and some details of the edifice were worked out soon after in an important paper, 'Hedges: a study in meaning criteria', by George Lakoff (1972, 1973). The paper is an ambitious fusion of mathematical logic and linguistics. The result of the 1972 version, presented to the Chicago Linguistics Society, was to import vagueness from the domain of logicians and philosophers to that of professional linguists.

Lakoff's paper is chiefly concerned with vagueness as it applies to category membership, and addresses the concern that 'natural language concepts' such as 'tall' lack sharply defined boundaries. He makes the point by reference to work by the psychologist Eleanor Rosch, who asked subjects to rank a number of creatures by the degree to which they matched a prototypical ideal of 'bird'. A well defined hierarchy emerged, in which robins were seen as typical birds, eagles less so, chickens somewhat less, followed by penguins, with bats hardly at all, and cows not at all. Lakoff gives a résumé of Zadeh/Goguen fuzzy theory, and concludes that:

> one need not throw up one's hands in despair when faced by the problems of vagueness and fuzziness. Fuzziness can be studied seriously within formal semantics [. . .] For me some of the most interesting questions are raised by the study of words whose meaning implicitly involves fuzziness – *words whose job is to make things fuzzier or less fuzzy*. I will refer to such words as 'hedges'. (1973, p. 471, emphasis added)[4]

It is clear from some of Lakoff's examples – e.g. 'sort of', 'in a manner of speaking' – that 'words' in the definition is intended to include phrases. Note also that words that 'make things [. . .] less fuzzy' are included by Lakoff in the category 'hedge'; examples include 'typical', 'definitely'.[5] Brown and Levinson (1987, p. 145) observe that this sense is an extension of the colloquial sense of 'hedge'; in fact, this sense will feature very little in my pragmatic analysis of mathematics talk.

Lakoff's paper belongs to the linguistic tradition of 'truth-conditional semantics', the purpose of which is to determine the conditions under which a sentence is 'true' (as opposed, conventionally, to 'false'). However, the 'formal semantics' which Lakoff has in mind entails specifying conditions under which vague propositions could be said to be true to *some extent*. That extent is measured on a continuum from 0 (perfectly false) to 1 (perfectly true). Lakoff describes a precise valuation of vague predicates and develops a corresponding truth-valuation of propositions through an exact mathematical calculus of truth-degrees – a task that had been set in train by Zadeh.

Truth-degrees need first to be assigned to a set of atomic statements such as 'Jack is tall', 'A penguin is a bird' and 'A rhombus is a sort of rectangle'. Once fixed, the valuation of any composite statement is determined in a precise and non-negotiable way. For example, if Jack is rich to degree 0.7 and handsome to degree 0.4, then 'Jack is rich and not handsome' is true to degree 0.6 precisely.[6]

Whilst much of Lakoff's paper is taken up with technical details in mathematical logic, he begins from and frequently returns to the issue of the meaning of vague language *in use*. My subsequent linguistic studies in pragmatics – how speakers and writers use language to achieve their practical purposes – grew out of the root source of this paper.

A Taxonomy of Hedges

Hedges can be usefully viewed as one of four basic types. This observation was initially made in a study (Prince et al., 1982) of paediatric clinicians, whose spoken language in case-conferences turned out to be unusually rich in hedging – about one hedge every 15 seconds. A hypothesis that these hedges would mark ethical more than clinical dilemmas turned out to be unfounded. The following representative examples of physician–physician talk (*ibid.*, p. 85) have an authentic ring to them:

> Well, I think he's, uh – I think he's always se – I still think he's seizing a – a little bit.

> There is evidence that's been presented that makes me think that it might be a little risky.

To elucidate the ways that different hedges work, I shall introduce and illustrate the four types by reference to this corpus of physician talk, and also to some intuitive language data.

The first major type of hedge – a *Shield* – is exemplified above by 'Well, I think . . .' and 'There is evidence that's been presented. . . .'. These indicate some uncertainty in the mind of the speaker in relation to some proposition. The marker (such as *I think that*) lies outside the proposition itself, which may be unequivocal. For example, the sentence

I:3.1 Maybe the pharmacy is still open[7]

invests all the vagueness in the speaker's uncertainty, as opposed to any degree of openness of the pharmacy. The speaker is asserting a proposition (call it S):

I:3.2 the pharmacy is still open (S)

S is thus made available to others, who may then (if they so wish) discuss whether or not it is true, and to act on it if, for example, they are in need of aspirin. The effect of the hedged assertion 'Maybe S' is to comment on the plausibility of

S without qualifying S itself. This is an important distinguishing feature of Shields in relation to mathematical discourse. For example, with reference to a Pythagorean triple (x, y, z) with $x^2 + y^2 = z^2$

1:3.3   I think that *x or y is a multiple of 3*

The italicized part is a mathematical sentence in the object language (here, a subset of mathematics called Number Theory), whereas the hedge phrase 'I think that' is in a meta-language (English). This difference of linguistic status can be made more visible in the form

1:3.4   I think that $[3 \mid x \text{ v } 3 \mid y]$

Such a hedge presents a mathematical assertion in the form of a conjecture, and implicitly invites comment on the conjecture.

With a little fine-tuning, Prince et al. subdivide Shields into two kinds. The first is termed a *Plausibility Shield*, typified by 'I think', 'probably' and 'maybe'. A Plausibility Shield 'implicates' (i.e. infers, by a mechanism to be discussed in the next chapter) a position held, a belief to be considered – as well as indicating some doubt that it will be fulfilled by events, or stand up to evidential scrutiny.

The second kind, an *Attribution Shield*, implicates some degree, or quality, of knowledge to a third party. A favourite Attribution Shield with the clinicians, with evident attendant suspicion, was 'According to the mother . . .'. An Attribution Shield may even fail (for whatever reason) to divulge the source or informant.[8]

The second major category of hedges (*Approximators*) includes 'about' and 'a little bit'. In distinction to Shields, these Approximator hedges are located inside the proposition itself. The effect is to modify (as opposed to comment on) the proposition, making it more vague. For example, from the Prince et al. corpus

Um, the baby's blood pressure on the ride over here was also *about, uh, something between forty and fifty* palpable. (1982, p. 87, emphasis added)

A subcategory of Approximators – called *Rounders* – consists of the standard adverbs of estimation, such as 'about', 'around' and 'approximately', which are commonplace in the domain of measurements, of quantitative data.

The second type of Approximator is called an *Adaptor*. These words or phrases, such as 'a little bit', 'somewhat', 'sort of', attach vagueness to nouns, verbs or adjectives associated with class membership. These Adaptors exemplify the hedges which are the subject of Lakoff's semantic work (1972), and the issue here is class membership.

Prince et al. summarize their analysis in the form of a binary tree (see Figure 1). In Chapter 6, I examine the particular purposes achieved by these categories of hedge in mathematics talk. In particular, I shall show that, on occasion, speakers use Approximators for Shield-like purposes.

Figure 1   A taxonomy of hedges (after Prince et al., 1982)

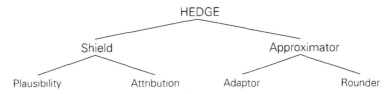

HEDGE

Shield   Approximator

Plausibility   Attribution   Adaptor   Rounder

Viewpoint: Philosophy of Language

The linguist Geoffrey Leech has acknowledged the contribution of philosophy to the pragmatic understanding of language:

> [when] linguistic pioneers such as Ross and Lakoff staked a claim on pragmatics in the late 1960s, they encountered there an indigenous breed of philosophers of language who had been quietly cultivating the territory for some time. (1983, p. 2)

Philosophers had made the problem of vagueness very much part of their territory from about 400 BC, in disputes and attempted resolutions concerning a type of paradox called 'sorites', which means 'the heap'.[9]

The first definition of vagueness is, in fact, due to the philosopher-mathematician Peirce, in a dictionary entry:

> A proposition is vague when there are possible states of things concerning which it is intrinsically uncertain whether, had they been contemplated by the speaker, he would have regarded them as excluded or allowed by the proposition. By intrinsically uncertain we mean not uncertain in consequence of any ignorance of the interpreter, but because the speaker's habits of language were indeterminate; so that one day he would regard the proposition as excluding, another as admitting, those states of things. Yet this must be understood to have reference to what might be *deduced* from a perfect knowledge of his state of mind; for it is precisely because those questions never did, or did not frequently, present themselves that his habit remained indeterminate. (1902, p. 748)

Peirce's definition is not easy to penetrate without reference to Peirce's semiotic theory, but can be seen to involve (1) modality, uncertainty concerning 'possible states of things', (2) the question of borderlines between what is and what is not of a given kind, and (3) inconsistency of speaker habit.

Peirce later theorizes about the nature of vagueness in his 1905 paper 'Issues of pragmaticism' (reprinted in Peirce, 1934), drawing a distinction between two kinds of indeterminacy; *generality* and *vagueness* (*ibid.*, para. 5.447). The indeterminacy of the former lies in the fact that it refers, not to this or to that, but to anything (in a given class). The indeterminacy of the latter has more to do with class boundaries, so that its field of reference is indeterminate. This distinction will become significant in Chapter 5, considering indeterminacy with regard to the referents of pronouns.

Peirce acknowledges the endemic presence of vagueness in everyday discourse:

> In another sense, honest people [. . .] intend to make the meaning of their words determinate [. . .] they intend to fix what is implied and what is not implied. They believe that they succeed in doing so, and if their chat is about the theory of numbers, perhaps they may. But the further their topics are from such precise, or 'abstract' subjects, the less possibility is there of such precision of speech. In so far as the implication is not determinate, it is usually left vague. (1934, para. 447)

Whereas Peirce is concerned only with vagueness and propositional meaning, my interest extends to propositional attitude and the social functions of vagueness. I will therefore want to go further than Peirce, to claim (e.g. in Chapter 8, case 8) that vagueness has an essential communicative function in 'chat [. . .] about the theory of numbers'.

Peirce is the originator of the philosophical position called 'pragmatism'. His successors include Dewey (1923), Rorty (1980) and Bernstein (1983), who refute objectivism and subscribe to a view that all knowing involves interpretation. The pragmatic tradition locates thought and enquiry in argument and the development of sound judgement (Giarelli, 1988, pp. 23–24). A fallibilist view of *mathematical* knowledge can clearly be seen to flow out of this philosophical stream (Ernest, 1991, p. 201).

Vagueness is also a recurrent theme in the philosophical work of another mathematician, Bertrand Russell, from 1913 to 1948. His main ideas are assembled in the 1923 paper 'Vagueness', in which he holds that vagueness is not inherent in 'things', but is a property of the symbols (including words) that represent them. Things are what they are; both vagueness and precision are features of their representation. He goes on to argue that all language is vague. For example, the word 'red' is vague because

> there are shades of colour concerning which we shall be in doubt whether to call them red or not, not because we are ignorant of the meaning of the word 'red', but because it is a word the extent of whose application is essentially doubtful. (1923, p. 85)

Russell makes an elaborate case for the vagueness of all words, including names and even logical connectives, and this lexical vagueness in turn infects all propositions. One vague word is enough to entail the vagueness of a sentence. Russell argues that all knowledge is vague, and our communication of knowledge by language is contaminated by vagueness. In adopting such an extreme position, it is as if Russell is engaging in a philosophical game. However, by exposing us to the thought that everything is vague, he raises our awareness of the possibility of vagueness when we may least expect it. Moreover, whilst asserting its inevitability, Russell explicitly acknowledges a positive epistemic characteristic of vagueness which will feature later in pragmatic analysis of mathematics talk:

It would be a great mistake to suppose that vague knowledge must be false. On the contrary, a vague belief has a much better chance of being true than a precise one, because there are more possible facts that would verify it. If I believe that so-and-so is tall, I am more likely to be right than if I believe that his height is between 6 ft. 2 in. and 6 ft. 3 in. (p. 91)

More recently, Tiegen (1990) has captured the same notion – which he calls the 'preciseness paradox' – in the following, similar, terms. Suppose two speakers, P and V, give similar information (the date of an historical event, say), but P is more precise than V (who is vague). One would then suppose P to be better informed than V in the field (history) in question. However, disregarding the knowledge level of the two speakers, V's statement is more likely to be true than P's, because the set of conditions (dates) which would make it true is much greater. The tendency is therefore to trust the person who is most likely to be wrong. Tiegen confirmed this prediction empirically.

The same point is made yet again by the Oxford philosopher of language, John Austin:[10]

And isn't it surprising that precision should be paired off with incorrigibility, vagueness with impossibility of verification? After all we speak of people 'taking refuge' in vagueness – the more precise you are, in general the more likely you are to be wrong, whereas you stand a good chance of not being wrong if you make it vague enough. (1962a, p. 125)

In a discussion of vagueness in *Sense and Sensibilia* (1962a), Austin recognizes that 'vague' covers a number of concepts. As he puts it:

'Vague' is itself vague. (p. 125)

Others have encountered the same dilemma.

Vagueness is not easy to characterise or define. One reason for this difficulty is that there appear to be a number of different conceptions of vagueness, and it is not clear just what they have in common. (Burns, 1991, p. 3)

If one looks more closely at this vagueness one soon discovers that the term itself is rather vague and ambiguous: the condition that it refers to is not a uniform feature [. . .] (Ullmann, 1962, p. 118)

The words which signify these aspects of vagueness are not themselves precise or uncontentious. Channell (1994, pp. 34–38) distinguishes between vagueness and ambiguity, whilst pointing out that an utterance can have both. Her main point is that vagueness is a much more significant factor than ambiguity in real communication, because ambiguity is usually automatically resolved by hearers, whereas vagueness 'often plays an important part in the act of meaning'.

Central to an understanding of vagueness in use is Austin's comment (1962a, p. 125) that vague features of language are: 'not necessarily defects, *that* depends on what is wanted'. His brief but important contribution to the discussion of vagueness affirms what Russell had proposed: that vagueness in language can be, in some circumstances, not a flaw but a 'Good Thing'. Peirce, in defining 'a proposition is vague when [...] it is intrinsically uncertain whether [the speaker] would have regarded [certain things] as excluded or allowed', is saying that vagueness offers speakers a way of saying something without needing to be sure of its scope of reference. From this perspective vagueness is not a limitation but a means to do things which are inhibited by precise communication. Allowed vague components of language, I can still say *something* without having to say something precisely. But, more than that, I can exploit the vagueness to convey something of my 'propositional attitude'; for example, I can let it be known that what I am saying is provisional.

Modality

Modality is a dimension of language which has some prominence in this book, because I am concerned with the ways in which speakers convey conviction, or the lack of it. From a semantic viewpoint, modality has to do with attitudes on the part of the speaker (or writer) towards the factual content of what s/he says. This is what I have referred to earlier as propositional attitude.Within a predominantly technical discussion, Michael Halliday (1976, p. 197) suddenly penetrates to the heart of the educational significance of modality:

> Modality is a form of participation by the speaker in the speech event. Through modality, the speaker associates with the thesis an indication of its status and validity in his own judgement. [...] This, we are suggesting, is not a minor or marginal element in language, but one of its three primary functions, that concerned with the establishment of social relations and with the participation of the individual in all kinds of personal interaction.

Modal *logic* distinguishes between propositions that are necessarily true and those that are contingently true. Necessity and possibility are the two aspects of *alethic* modality. Propositions which are necessary truths are referred to as alethic necessities, those which are not necessarily false as alethic possibilities.

A second kind of modality – *epistemic* – is concerned, not so much with objective truth as with human knowledge and belief. For example, the epistemic sense of 'It may be raining' would be 'I have reason to entertain the possibility that it is raining'. The epistemic quality is explicit and clear in 'I think that a bus will come soon'. Epistemic modality enables the speaker to indicate her/his commitment to the truth of a proposition. Epistemic modals are included in 'the general category of means used to convey the attitude of the speaker towards the utterance he makes' (Dubois, 1969, p. 118). An epistemic modal continuum has confidence and doubt at its extremes.

A third kind of modality – *deontic* – is related to the necessity and possibility of action. The appropriate notions here are obligation and permission. Thus: obligation = necessity to act, permission = possibility to act (Stephany, 1986, p. 376).

> I:3.5   You must return the book to the library.

> I:3.6   You may borrow the book until Friday.

Halliday (1976, p. 199) carefully refers to this kind of language as 'modulation' rather than modality, and observes that here the auxiliaries (*must* and *may* in my invented examples) have nothing to do with speaker attitude to propositional content; rather, they actually modulate the content itself – they are part of the ideational (i.e. content-expressing) meaning of the clause in which they are positioned. Note the parallel with the distinction drawn by Prince et al. (1982) between Shields and Approximators, so that Shields effect modality, whereas Approximators effect modulation.

In English, modality is achieved syntactically in one of three possible ways:

- usually by the use of modal auxiliary verbs such as 'may', 'can', 'must', 'could';
- also, by the use of epistemic adverbs such as 'possibly', 'maybe', 'perhaps', 'obviously';[11]
- less commonly, by the use of verb moods and tenses. For example, 'She appeared as though she were asleep' uses the 'modal preterite', i.e. past tense.

In a study of the modal auxiliaries 'must', 'should', 'ought', 'may', 'might', 'can', 'could', 'would', 'will', 'shall', Coates (1983) emphasizes the importance of epistemic modality in normal (as opposed to logical) language (p. 18), and also the essentially subjective nature of epistemic modality (pp. 18–20), which serves to indicate the speaker's confidence (or doubt) in the truth of the proposition s/he expresses. Coates uses the term 'root' for some types of non-epistemic (including deontic) modality. This root category is problematic if one accepts Stubbs's position that

> [. . .] all utterances express not only content, but also the speaker's attitude towards that content. (1986, p. 15)

In effect, Stubbs declares that there is no root modality, that no utterance is neutral with respect to the speaker's commitment to what s/he is saying. Yet there is a strong case for believing that there is a root quality in mathematics – and that it is highly valued. Recall the words of Hardy:

> 317 is a prime, not because we think so [. . .] but because *it is so*, because mathematical reality is built that way. (1940, p. 130)

Hardy's passionate confidence is certainly epistemic, but when the commentary is stripped away from his uttererance, what remains is

317 is a prime,

which simply asserts what is the case. Surely many young men and women desert mathematics for literature because mathematics is *so numbingly root*. In their experience, not even their teachers seem to *care* much either way about its ideas and theorems; there is no passion in their experience of it, neither confidence nor doubt, no commitment of any kind.

In the end, I have found a working distinction between root and epistemic to be helpful in relation to utterances of a mathematical kind. The distinction is not clear-cut, any more than one should expect in a book dedicated to vagueness. I shall speak of epistemic modality in connection with commitment (or the lack of it) to hypothetical states of affairs, as opposed to perceived actualities, and particularly in connection with the attitude of the speaker to what s/he asserts about such contingent or hypothetical matters. Root modality will be (more or less following Coates) a more matter-of-fact kind of modality, not caught up in the belief or commitment of the speaker. It will soon be apparent that epistemic modality is a prevalent and important means by which pupils mark tentativeness about their mathematical assertions.

Pragmatics and Vagueness

Peirce's philosophy of pragmatism recognizes the place of intention and interpretation in the determination of meaning. Pragmatics considers language from the point of view of the user – choices, constraints, purposes, and so on. To some extent, pragmatics has arisen in response to the limitations and artificial abstraction of truth-conditional semantics.

Pragmatics is a young linguistic discipline, and has suffered to some extent from a reputation as the 'waste-basket' of linguistics[12] – to which to consign linguistic matters which were not the concern of 'pure' syntax or of truth-conditional semantics. However, the discipline of pragmatics is now richly endowed theoretically, and the 'waste-basket' reputation is manifestly *passé*. Both Levinson (1983) and Mey (1993) explore at length what it is that characterizes pragmatics and how it complements syntax and semantics. Mey summarizes:

> Pragmatics is the science of language seen in relation to its users. That is to say, not the science of language in its own right, or the science of language as seen and studied by the linguists [. . .] but the science of language as used by real, live people, for their own purposes and within their limitations and affordances. (1993, p. 5)

Now Lakoff's resolution of the problem of vagueness by means of (fuzzy) formal semantics attempts, by precise truth-valuation, to take the uncertainty out of vagueness – which is to ignore when and why, in the world, anyone should want to be vague in the first place. Or how communication can be possible, let alone effective, when (as Russell suggests) it is infused with vagueness. Such questions are outside the scope of truth-conditional semantics, but well within the province of pragmatics.

Lakoff's definition of 'hedges', '*words whose job is to make things fuzzier or less fuzzy*', conveys the presupposition that language can *do* things. Lakoff (1973, p. 490) and others (Prince et al., 1982) include as hedges certain epistemic modal forms such as 'I think' and 'maybe' which obscure the degree of commitment of the speaker to what s/he is saying. Stubbs (1986) associates this with inexplicitness, 'which implies vagueness and therefore deniability'.

Concern for speaker intention is thus seen to emerge as an issue more salient than truth conditions. In 1977 Sadock offered a radical pragmatic alternative to Lakoff's precise semantic truth-valuation of sentences containing hedges:

> it is the *purpose* of the estimate that essentially determines how close to the truth it must be to be warranted. (1977, p. 434, emphasis added)

Sadock argues that:

> the role of an approximator [. . .] is to *trivialise the semantics* of a sentence, to make it almost unfalsifiable. (p. 437, emphasis added)

Extensive study of the purposes underlying vague language, including hedges, has been undertaken by Channell, who identifies (from empirical data) a number of goals which speakers and writers achieve by the use of vague expressions. These include:

- giving the right amount of information;
- deliberately withholding information;
- saying what you don't know how to say;
- covering for lack of specific information;
- acknowledging and achieving an informal atmosphere;
- expressing uncertainty;
- downgrading the importance of something so as to highlight something else;
- expressing politeness, especially deference;
- protecting oneself against making mistakes.
  (Channell, 1985, 1990, 1994)

Many of these goals are evident in mathematical conversations. Channell's identification of these purposes presents an extremely useful pragmatic starting point from which to consider vagueness in such conversations. For example, it is not at all uncommon for novice speakers of mathematics to have difficulty in giving expression to their mathematical thoughts, perhaps because they lack fluency in the mathematics register. In Chapter 5, I shall show some ways in which vague language enables such pupils to say what they don't know how to say. Similarly, uncertainty and lack of specific information are ever-present epistemic factors in any creative discussion of mathematics, where conjectures are asserted in advance of certain knowledge. The last of Channell's goals – protecting oneself against making mistakes – is associated with Sadock's notion of making a sentence 'almost unfalsifiable'. The use of hedges to introduce vagueness into propositions in mathematics talk will be examined in detail in Chapter 6.

Summary

The world, in so far as it is entertained by thought or expressed in language, is infused with vagueness. This may be perceived as problematic if precision is, or is deemed to be, desirable, and is inhibited by vagueness. The problem of vagueness is redeemable only, if at all, by the imposition of artificial syntax and interpretation, to achieve precision of expression and meaning in a stipulative way. Even vague language itself can be furnished with precise interpretation in such a way, recognizing that this amounts to a formal game of some kind, irrespective of relevance and truth. Mathematics itself, however, is seen to be fallible, shot through with uncertainty as to the origin and the truth of its propositions. The learning of mathematics may be viewed as a process of active construction in which, from time to time, vagueness is experienced individually and expressed socially. The enthymematic nature of inductive reasoning suggests that knowledge arrived at by inductive inference is provisional, plausible rather than certain.

This social perspective on learning presents an entirely different *pragmatic* perspective on the phenomenon of vagueness as a component of mathematical discourse. It will become apparent that vagueness can be viewed and presented, not as a disabling feature of language, but as a subtle and versatile instrument which speakers can and do deploy to make mathematical assertions with as much precision, accuracy or as much confidence as they judge is warranted by the *circumstances* of their utterances as well as by their content.

Vagueness turns out to be a unifying theme, common to many of the pragmatic aspects of language which I identified in mathematics talk in general, and in conjecturing talk in particular. Although strict definition and circumscription of vagueness are problematic, it is useful to observe that:

- vagueness complicates the truth-conditional semantics of the propositional content of language; whilst
- pragmatics deals with aspects of meaning that are outside the scope of truth-conditional semantics.

A principle which guides my choice of subject matter is the aim to expose some of the ways that participants in mathematics talk use vague language for interactional and transactional purposes. My analysis occasionally extends beyond a strict focus on vagueness, in pursuit of insight into 'the extremely subtle pragmatic interpretive judgements regularly made by both teachers and pupils in the course of mathematics teaching and learning' (Pimm, 1994, p. 167)

Notes

1   The point is actually more subtle than Freudenthal suggests, in that the encyclopaedia entry is guilty not of inappropriate precision but of non-standard use of 'Rounders' – see Chapter 6. Freudenthal's intended point was nicely made in a gardening feature some years ago in the *Cambridge Evening News*, presumably 'updated' from a pre-metric source. Describing how to build some structure or other it advised: 'Use a piece of wood about 7.62 cm wide.' More recently a home-brew kit has instructed me to 'add 907 grams (2 lb) of white, granulated sugar'.

2   This view derives, of course, from the philosophy of Plato. The use of the name 'platonism' to describe a philosophy of mathematics is due to Paul Bernays (1935).

3   These ideas are now in widespread use in the design of all kinds of electronic systems, notably control systems for everything imaginable, including drum voltage on photocopiers, anti-lock brakes, power and cooking strategy on microwave ovens (Kosko, 1994). Essentially, in the case of air conditioners, it works as follows. X, the system input, is temperature. X has five overlapping fuzzy subsets COLD, COOL, OK, WARM and HOT. Five fuzzy membership curves indicate (in a provisional way, subject to adjustment) the extent to which any temperature belongs to each subset. Apart from the extreme subsets, the curves are symmetrical. Y, the output, is the speed of the system motor. Y, too, is separated (not partitioned) into five fuzzy sets, STOP, SLOW, MEDIUM, FAST, BLAST. Again, five fuzzy membership curves indicate the extent to which any speed belongs to each subset. The connection between X and Y is fixed by some discrete pragmatic rules; basically, the hotter it is, the faster the motor speed needs to be. But these are fuzzy rules; for example, 'if warm, then fast'. The input from the temperature sensor, together with the fuzzy membership graphs and fuzzy set algebra, lead to a simple computation of the motor speed. The system can be fine-tuned, in a pragmatic way, to meet the preferences of the user.

4   Taken out of the context from which it arose, Lakoff's choice of the word 'fuzzy' in his definition of 'hedge' seems somewhat self-consciously colloquial in a careful and rigorous academic paper. The choice is, without doubt, a tribute to Zadeh, who chose 'fuzzy' in preference to the philosophically standard 'vague'. Along the way Zadeh considered and rejected 'cloudy' (Kosko, 1994, p. 145). Evidently, in the definition of 'hedge', 'fuzzy' may be taken as synonymous with 'vague'.

5   An example which comes to mind, deeply embedded in the language of Number Theory, is 'perfect square', meaning the square of an integer. Since in non-analytical Number Theory one is dealing only with integers anyway, the hedge 'perfect' is superfluous. In fact its inclusion regularly confuses students, who ask me whether a perfect square must also be a perfect number (equal to the sum of its proper divisors); in which case there would be no 'perfect' squares . . .

6   For let j denote Jack and let R, H be the sets of Rich and Handsome people respectively. Then, given that $m(R, j) = 0.7$ and $m(H, j) = 0.4$, it follows that $m(R', j) = 1 - m(R, j) = 0.3$. Hence $m(R' \cup H) = \max\{0.3, 0.4\} = 0.4$. Now $R \cap H' = (R' \cup H)'$, and so $m(R \cap H') = 1 - (R' \cup H) = 1 - 0.4 = 0.6$.

7   The reference I:*n.m* indicates that, whilst the suggested utterance is intended to be plausible, it is invented (the *m*th such in Chapter *n*) for the purpose of exposition or clarification. The issue of the validity of such intuitive data is discussed later in the next chapter.

8   For example, as in the following notice (I have italicized the hedge preface) published in March 1996 by the Library Syndicate of the University of Cambridge: 'The Library

Syndicate is concerned at the increase in the number of cars parked outside the design-ated spaces at the front and side of the University Library. [. . .] *There is evidence to suggest that* the shortage of spaces is exacerbated by people using the Library car park when they are not in the Library.'

9 The paradox poses the question: if single grains are removed one at a time from a heap of sand, at what point is it no longer a heap? An alternative formulation concerns a hairy man losing single hairs from his head; when is the man bald? The sorites paradox is important in so far as it appears to have identified, for Western thought, a philosoph-ical issue which potentially permeates all of life. A resurgence of interest in the sorites in modern times is evident (Russell, 1923; Goguen, 1969; Rolf, 1981; Sperber and Wilson, 1986b; Burns, 1991; Kosko, 1994; Williamson, 1994).

10 Austin is best known as the originator of the theory of speech acts, which I describe in the next chapter.

11 Halliday (1976, p. 193) also lists variants on the adverbial forms.

12 The expression derives from Bar Hillel's caveat (1971, p. 405) 'Be careful with forcing bits and pieces you find in the pragmatic wastebasket into your favorite syntactico-semantic theory. It would perhaps be preferable to first bring some order into the contents of this wastebasket.'

4   Discourse and Interpretation

Interpreting an utterance is ultimately a matter of guesswork, or (to use a more dignified term) hypothesis formation. (Leech, 1983, pp. 30–31)

This chapter prepares for the analysis of transcript data in the following three chapters. The purposes of that analysis will be to:

- infer the transactional meanings of many of the utterances in the conversations, and in particular the way that individual children structure aspects of their mathematical understanding;
- investigate how participants used and decoded language to support their contribution to a co-operative social interaction;
- consider the motives which determine the character of that contribution to that mathematical interaction.

The pursuit of each of these purposes is an *interpretive* process. Whilst Leech's dictum (above) is superficially cynical, it would be optimistic to insist that it was entirely false. Nevertheless, I suggest that Leech's choice of the word 'guesswork' is unfortunate. The responsible analyst will take what steps s/he can to minimize guesswork by drawing on what is known about the process of analysis. In my own interpretive accounts of mathematics talk, I shall draw on the literature of linguistics – of pragmatics, in particular – with an eye to *regularities* and *purposes*.

- Where discourse linguists have found distinct ways of describing and analysing aspects of language (in particular, in terms of indirect speech acts, implicatures and preference organization) to be of interactive significance, I shall use these approaches to inform the analysis of my data. In this case, the data are overlaid with particular mathematical significance – the mathematical component of the context will be one that will also inform my interpretation.
- Where linguists have associated particular types of language (Channell's work on vague language is the paradigm example) with particular *goals*, I shall look for those goals in situations where such language appears in mathematics talk.

The interpretation of the meanings and motives of others is, first and foremost, a synthetic act of meaning-making for the analyst, whose 'reading' of a particular utterance must be made to fit, to be consistent with, the way that s/he construes the utterance in its multidimensional context – social, psychological, mathematical, textual and who knows what else. The participants in the conversation are bound to be making such interpretations 'on the hoof' within the conversation itself, otherwise they could make no interactive contribution to it. The analyst after the event (in contrast with the clinical analyst-in-conversation) is subject to no interactive obligations, and so may 'interpret' at her/his leisure. At the same time, s/he has no interactive opportunity, and so is denied the possibility of testing interpretive conjectures in order to validate or refute them (Steffe, 1991, p. 178).

I proceed now to survey some (mainly pragmatic) aspects of language, to which I shall refer in subsequent chapters. The survey is organized into two parts. The first examines the topic of reference, especially the role of pronouns. The second considers some approaches to the analysis of discourse; drawing in particular on speech act theory, conversational implicature, politeness theory and conversation analysis.

It will become clear in subsequent chapters how these issues relate to my analysis of mathematics talk. They will be seen to provide certain cues and expectations about the interactions, to inform interpretation of the interlocutors' meanings and motives.

Reference

The theme of Chapter 5 is the value of vagueness as a means of referring to 'things' in mathematics talk. The linguist John Lyons, presenting a traditional view of reference (1968, p. 404), indicates that words refer to things. Thus, in the sentence 'The book is blue', the definite noun phrase 'the book' is held to refer to some real-world object (a book). In a subsequent exposition, however, Lyons (1977, p. 177) holds that it is the *speaker* who refers, by the use of some expression. This is a pragmatic, as opposed to lexical semantic, view of reference, which is viewed as an *action* on the part of the speaker/writer (Brown and Yule, 1983, p. 28). Successful reference is achieved by the speaker if the hearer (as interpreter of the utterance) is able to recover the intended referent(s).

From a constructivist perspective, Glasersfeld (1995) adopts a more complex, interactional approach to reference. He asserts that it is naive to suppose that words refer to independent objects, or that a universal meaning is shared by independent speakers. He argues that the function of words is to evoke or generate 're-presentations' of experience (actual or imaginative). Conversely, experiences may call up, for an individual, the words which s/he associates with those experiences. The essence of successful reference is not sameness, but compatibility. Each language user co-ordinates a word-experience pair;[1] linguistic communication then consists of co-ordinations which are *compatible* rather than identical. A well known instance of this phenomenon relates to colour. Suppose, never having seen me, you arranged to meet me at Arrivals at Heathrow airport, and I told you that I would be wearing a green jacket. Suppose then that you greeted me as I emerged into the public space. The success of the arrangement depends only on the pragmatic compatibility, for the purpose of our rendezvous, of our two word-experience co-ordinations (of 'green' and of 'greenness'); it is not possible to infer that we share a common understanding of 'greenness'. Diagnostic mathematics talk is useful in so far as it explores and reveals the extent of the compatibility of the teacher's co-ordinations and the pupil's, within the domain of discussion. Talk about rectangles, for example, or about multiples, may (or may not) expose incompatibilities that necessitate renegotiation – or correction – of word-concept co-ordinations, depending on the circumstances. Very similar notions of reference to these explicitly underpin the analysis of mathematics classroom interactions in Cobb et al. (1992). Their account rests on terms such as intersubjectivity, and meanings which are consensual or taken-as-shared.

Some precision of reference (a high level of compatibility, from the social interactionist perspective) is achieved by the use of names (e.g. 'Windsor Castle') and noun phrases (e.g. 'the man at the bus stop'). The referents of pronouns are *potentially* vague, in the sense of ambiguous or indeterminate, if they cannot easily be associated with a co-referential name or noun phrase. Context then becomes crucial for the purpose of interpretation.

Speakers depend on this, for example to achieve economy of utterance. In interpreting the question (father to son) 'Have you fed the cat?' the son might be expected to presuppose that his father was referring to their household cat, rather than to someone else's. When the son replies, 'Yes, I fed her this morning,' the pronoun 'her' is expected to be understood by the father as co-referential with his earlier 'the cat'. Hence the father would infer that, in using the pronoun, his son was referring to the household cat.

Pronouns and Reference: Power and Solidarity

The intended referent of a pronoun may be associated with a name or noun phrase elsewhere in the text. Thus in the sentence

I:4.1 The next candidate to be interviewed was well-qualified for the job, and the panel agreed that she was impressive.

both the pronoun 'she' and 'the next candidate to be interviewed' refer to the same person (that is to say, the speaker uses both to refer to the same person). The use of the pronoun is said to be anaphoric, being co-referential with a noun phrase which was uttered earlier in the discourse, in this case earlier in the same sentence. Of course, pronouns may refer forwards ('cataphora') as well as backwards, as in the demonstrative

I:4.2 **This** will be the last time I shall ask you to help me.

In the case of both anaphora and cataphora, the pronoun clearly relates to another item or items in the text or utterance; use of the pronoun is economical (Leech et al., 1982, p. 191), and lends interest by avoiding the tedium of repetition of a noun or noun phrase.

It is not unusual for speakers to use pronouns in an irregular, somewhat anarchic way, with the pragmatic effect of conveying a range of social dimensions and attitudes to themselves and their audience:

> When I got there [Oxford], **I** think the first thing **I** learned was that for the first time in **my** life **you** were totally divorced from **your** background. **You** go there as an individual. So what did **we** learn? (Margaret Thatcher, in a television interview, 29 March 1983, quoted in Rees, 1983)

In this example the speaker uses singular and plural, first and second person pronouns co-referentially. There is a fascinating shift from very personal recollection ('I') to description of shared experience conveyed in the second person ('you'), to reflection with hindsight ('we'). Knowledge of this particular speaker might prompt the suspicion that the 'we' is the so-called royal plural, a pronoun used by the (then) Prime Minister to refer to herself. The pronouns code and convey aspects of speakers' personal identity and group association.

These pronominal variations can be associated with delicate shifts of social positioning of the speaker in relation to his/her audience. Essentially, pronouns may be used to convey messages of power over or solidarity with others. Brown and Gilman (1970) give a classic account of the power of pronouns as linguistic devices for expressing social distinctions in non-British European languages. French, for example, retains a pronominal social semantics in its 'T-V' (*tu, vous*) system: *tu* is the marked singular, expressing intimacy or certainly informality, occasionally condescension (to children by default). *Vous*, the unmarked plural, is also the singular for public or formal conversation, occasionally used as a marked pronoun of respect (or distancing with adults).

With the decline and disappearance of 'thou', English is almost unique among Western European languages in having neither plural nor honorific distinction in second person pronouns.[2] The complexities and instabilities of the functioning of this system in seventeenth-century English are clearly and succinctly illuminated by Hodge and Kress (1988, pp. 41–45) by reference to Shakespeare's *King Lear* and Thomas Middleton's *A Chaste Maid in Cheapside*.

In the context of pedagogy, the schoolteacher's 'we' may, at different times, convey extremes of distance and mutuality. I shall return to this theme in the next chapter.

Referents of 'You'

The use of 'you' to refer to generalities is familiar in mundane language use, but has attracted little analysis in the literature. This is fertile territory for mathematics education.

Helen Simons (1981, p. 39) records the following fragment of an interview with a 15-year-old girl. The subject was participation in class discussions.

> *HS:* Did you feel . . . that you did have things to say?
> *P:* Yes. But often other people said them . . .

HS: And that put you off saying something another time, did it?

P: Umm. If you say something you sometimes think that if you say something wrong people are going to think it is funny.

Simons uses 'you' to refer to the girl. The informal way that the pupil uses 'you' is utterly familiar. She refers, not to Simons, but (presumably) to herself and other pupils. Simons makes no comment on this – whilst it would look out of place in a formal text, it is a perfectly acceptable use of 'you' in speech, of the following kind:

> The pronoun of the second person may be used vaguely to denote someone (often the speaker himself) to whom something happens, or may happen, in the ordinary course of events:
>
> > It was not a bad life. You got up at seven, had breakfast, went for a walk, and at nine o'clock you sat down to work.
>
> (Zandvoort, 1965, p. 128)

Now this kind of 'you' is vague in that it is unclear who is meant to be included by it, the only certainty being the speaker 'himself' (by implication, since s/he does not choose a third person pronoun). This exemplifies the kind of reference indeterminacy associated with generality (Chapter 2). The fragment of dialogue which follows was broadcast on Radio 4 on 24 February 1995. Sue Lawley, host of *Desert Island Discs*, interviews Jimmy Knapp, President of the TUC:

Lawley: Are **you** an emotional man? Can **you** be moved by music?

Knapp: Aye, I think **you** can. There's a stirring in the breast that **you** can't deny.

Lawley's questions are clearly addressed directly and personally to Knapp. He is, after all, her chat-show guest, and she wants to know what kind of a man he is. The potential ambiguity of her 'you' is demonstrated by his response, in which he deflects the spotlight from himself to comment – with uses of 'you' of the kind indicated by Zandvoort – on the general effect of music on human emotions. Knapp's 'you' is deliberately *impersonal*, and vague as to who is comprehended by it. This ambiguity was mischievously exploited in the *Clive James Show* (Carlton Television ITV, 16 July 1995). Addressing an image of Margaret Thatcher on a giant video screen, Clive James, with a glint in his eye, suggests that there was once something between them, and asks the image:

James: What do you suppose that did to me, an impressionable young man?

to which the pre-recorded and mischievously out of context image replies:

Thatcher: I think it gave you a flavour of what life was all about.

My supposition is that Thatcher's 'you' was originally a reference to some shared experience, as in the earlier quotation about life at Oxford. James exploits the ambiguity of her 'you' to suggest that she is addressing him. The type of marked pronoun alternation, which I noted earlier in Thatcher's reminiscences, is deployed for effect in literature. In one of his novels R. F. Delderfield gives this portrait of Evan Rhys-Jones, bank manager and landlord, described by his bank clerk and lodger, Charlie Pritchard.

> He had . . . a gravity that **you** could mistake for dignity until **you** adjusted to the maddening deliberation of his movements. It was this characteristic that fascinated **me** on that first occasion, so that **I** found myself wondering how long it would take him to select a stick of celery, bring it up to his chubby jaws and produce the soft, carefully modulated snap, in contrast to his wife's regular volleys from across the table. **You** had the feeling that if **you** asked him to pass the salt the meal would falter to an uncertain halt, so in the end **I** compromised, watching him but listening to his wife's coy exploration of **my** non-existent love-life. (Delderfield, 1969, p. 13)

This alternation between first and second person pronouns, 'I/me/my' and 'you', has the effect of distinguishing experiences and feelings from detached observation and generalized objective comment.

In effect, 'you' is being deployed in place of the more formal indefinite pronoun 'one', which might be regarded as somewhat affected in English speech. There is some indication that there is a corresponding trend in French. Laberge and Sankoff (1980, p. 271), in a study of Montreal French, remark that '*Tu* and *vous* [. . .] are now locked in combat with *on* for indefinite champion, a title *on* thought it had locked up'. The issue is that of *generality* in relation to that which is being asserted.

> a detailed study of the contexts of use of indefinite *on* shows that *tu* and *vous* can be used in virtually all of them. Perhaps the most central element unifying these various contexts is the theme of generality or generalization. [. . .] It is important to note that the indefinite referent here is always vague as to the possible inclusion of speaker and hearer: *anybody* 'means' just that – possibly you, possibly me, or anyone else in like circumstances. (*ibid.*, p. 275)

The same linguistic trait is very evident in the Fawcett corpus of child language (Fawcett and Perkins, 1980). The many examples from the speech of 10-year-olds include a number of 'rules' of games:

> [Rules of 'Monopoly'] You buy lots of property and houses and then when someone comes along they land on you collect money that way as well.
>
> [Rules of bagatelle] You roll a ball down a chute 'n' it's got to go in one of the holes . . .

As I remarked, this matter seems to have escaped analytical attention with regard to English speech. I take it up again in the next chapter.

Deixis

The link between the use of pronouns and the pragmatic notion of deixis will be important in Chapter 5. Deixis is concerned with aspects of meaning that are inaccessible without the provision of context – for example, the use of a word or phrase whose referent is determined by the context of its utterance. Deictic features of speech and writing correspond to what philosophers call 'indexicals', which reveal attributes of place or person. Deictic forms such as 'you', 'now', 'here' are effectively context-dependent variables, or 'shifters' (Mey, 1993, p. 90). Divorced from the context of utterance, their meaning may be ambiguous or obscure. The authors of the ATM booklet *Language and Mathematics* (1980) refer implicitly to the communicative function of deixis:

> Everyday conversation is easy to understand, its meanings are clear, because we speak in the context of everyday. When we are in a bus shelter, and a bus comes round the corner, the words 'Here it is' have a clear meaning. (p. 17)

Similarly, as an example of temporal deixis, observe the difference in the intended immediacy of 'now' in the two utterances below and how the difference is clarified by the context:

I:4.3   I'm going for lunch now. [Context: workplace]

I:4.4   I suggest you begin the next chapter now. [Context: supervisor to student]

The word 'deixis' is usefully related to its Greek root *deiknumi*, meaning 'to show' or 'to point'. The familiar noun 'paradigm' derives from the same root – to mean an example which acts as a pointer to a general type. The Greek word also means 'to prove'. A *diknumi* proof (Fauvel, 1987, p. 5) is one which is presented – typically by means of a diagram of some sort – in such a way that no explanation is necessary, for one can 'see' the result and the argument. For example, a suitable arrangement of pebbles in pairs 'demonstrates' that the sum of two odd numbers is even. The example displayed – the arrangement of a particular set of pebbles – is a generic example of the kind which I discussed in Chapter 2, in that it *points* to a more general truth. In the *diknumi* proof, a train of thought is shared yet unspoken.

Mühlhäusler and Harré (1990) emphasize the primacy, in their view, of the deictic role of pronouns.

> From developmental evidence we know that the ability to use pronouns in their deictic function predates the correct use of pronouns in their anaphoric function. [. . .] whilst anaphoric pronouns almost invariably contain deictic information, the reverse is not true. One is led by such observations to conclude that *the deictic function of pronouns is their primary function* . . . (p. 58, my emphasis[3])

Roger Wales, in a survey of developmental aspects of deixis (1986), observes that:

[deictic expressions] serve as a meeting point for semantic, syntactic and pragmatic aspects of language. This is because they are, to use G. Stern's (1964) term,
contingent expressions. By this is meant that, to interpret them, the interpreter
needs not only context-independent semantic information but also information
which is contingent on the actual (or construed) context. They are used to direct
the hearer of a communication towards some object or event. (p. 401)

Deixis is concerned with ways in which language draws on and points to context. I
shall show how deictic elements of mathematics talk can enable students to convey
important ideas, especially when they lack the necessary vocabulary and fluency in
the mathematics register.

Some Approaches to Discourse

Over the last forty years or so a number of approaches to the analysis of discourse
have arisen and evolved. The roots of these analytical traditions have been in
disciplines such as philosophy, sociology and anthropology, and their particular
emphases and contributions vary accordingly. I now consider some of these approaches, each of which offers some insight into interactive dialogue. My purpose
is not to suggest which of the frameworks is best or most appropriate, but (in the
spirit of Schiffrin, 1994) to make a number of approaches available for my own
analysis of mathematics discourse.

The first approach is based on Austin's insight that an utterance can be a
means of performing an action.

Speech Acts

An account of how speakers 'do things with words' can be seen as an outcome of
the theory of 'speech acts', which, for three decades, has occupied a central place in
pragmatics. A declarative utterance such as 'The window is open' expresses a
proposition with a truth-semantic value – true or false. By contrast, the imperative
utterance 'Shut the window' does not express a proposition in the truth-conditional
sense, since it cannot, under any conditions, be evaluated as true or false, or even
something in between. It is an order, requiring the hearer to do something; an action
performed by language (speech in this case).

A quite different example of such an action is 'Good luck!', which is a wish
(*may you have . . .*), a projection of a felicitous state of affairs in the near future.
I can assert the truth of my sincerity in making the wish, but not the truth of the
wish itself.

This insight is due to Austin (1962b), who called non-propositional requests,
wishes and the like 'speech acts'. The essential property of speech acts is that they
do something in the world, that they bring about (or have the potential to bring
about) a change in some state of affairs. They are 'performative' utterances.

A speech act is accomplished, canonically, by the explicit use of a performative or speech act verb (SAV) such as 'promise'. The non-necessity of an explicit SAV in a speech act is borne out by the 'Good luck' example. The formal (if somewhat odd) paraphrase 'I (hereby) wish you good luck' makes the SAV explicit.

Certain paradigm performatives powerfully convey the character of speech acts. When a priest says to the infant cradled in his or her arms, 'I baptize thee in the name of the Father', etc., then, as Mey (1993, p. 112) says, 'there will be one more Christian among the living'. The standard bench test for a performative verb is whether the adverb 'hereby' can be sensibly, even if unnaturally, inserted into the utterance containing it. Thus 'I hereby request you to shut the window' stands up to the test, whereas 'I hereby go to work by car' does not. Austin's initial position concerning speech acts was that they had to be associated with (possibly implicit) SAVs. He therefore urged the value of compiling a list of explicit performative verbs, a task that he judged would be 'a matter of prolonged fieldwork' – by which he meant using the 'hereby' test to uncover the SAVs in a dictionary.

Force and Felicity

A speech act has three different kinds of 'forces' (Austin 1962b), as follows:

- *locutionary* force: the actual act of speaking;
- *illocutionary* force: the direct, conventional action of making a promise, request, command, denial, etc.;
- *perlocutionary* force (or effect): the indirect (and sometimes unpredictable) consequences of the speech act which arise from the circumstances of, and beyond, its utterance.

Example: 'For tonight's homework I want you to finish the exercise.' Illocutionary force – an order to finish the exercise before the next lesson. Perlocutionary effect – three pupils are very late to bed that evening.

Whereas declarative utterances typically have truth-conditions, speech acts must satisfy certain 'felicity' conditions in order to count as an action. When the priest says to the child, 'I baptize thee,' his or her pronouncement is felicitous provided that the sacrament is properly convened and the priest is invested with the authority to perform it. In the case of orders and requests, the felicity conditions include the 'preparatory' requirement (Levinson, 1983, p. 239) that the speaker believes that the hearer is capable of carrying out the indicated action, that they would not necessarily do it without being asked; and also a 'sincerity' condition, that the speaker actually wants the hearer to do what s/he is being asked to do. If I ask you to close the door, the request is felicitous if the door is open, if I desire it to be closed, and if your situation is such that you are capable of closing it. On the other hand, a promise is not performative if I have no intention of honouring it; nor is a £1 million bequest, if I haven't got that kind of money.

Implicit Performatives

Human rites, particularly religious and legal ones such as baptism by a priest or sentencing by a judge, are carefully formulated to include (if not deliberately) performative verbs which emphasize the fact that those who speak the words are doing something in the world. When the priest says, 'I [hereby] pronounce you man and wife,' this is the precise moment in which the speech act effects the union of two persons in marriage.

Consider an earlier statement, in the same ceremony, 'I, Jack, take thee, Jill, [. . .] in sickness and in health.' At one level, this is a description of something that is true at that moment, with no performative marking. But the importance of these 'solemnized' words is that they constitute a commitment (for example, to mutual fidelity) and belief in a present and future state of affairs. Likewise, the verb 'to love' does not pass the 'hereby' test (? 'I hereby love you') and so does not qualify, in Austin's strict sense, as a performative verb. Yet (I would argue) the declaration 'I love you' simply cannot be uttered by the speaker as a pure and simple proposition, true or false: it has to convey an attitude, a commitment, a promise even, and in that sense it is a speech act.

Herein lies a difficulty for classical speech act theory, that a great many propositions can legitimately be viewed as 'acts' if the explicit performative requirement is relaxed. Austin himself recognized the difficulty and progressively adopted a broader view of speech acts. Levinson (1983) summarizes thus:

> what starts off as a theory about some special and peculiar utterances – performatives
> – ends up as a general theory that pertains to all kinds of utterances. (p. 231)

This is the position of Austin's student and successor (as regards speech act theory) John Searle (1969), that every utterance can be classified as some kind of speech act: for example, the proposition 'I am a man' is at the same time the speech act 'I [hereby] assert that I am a man'.

Indirect Speech Acts

The actual form of a speech act is not an arbitrary matter. Indeed, a whole range of personal factors – strategies, attitudes, beliefs, position relative to other persons, desires, commitments and detachments – may be encoded in the way that a speech act is formulated. In particular, 'indirect speech acts' are 'cases in which one illocutionary act is performed indirectly by way of performing another' (Searle, 1975, p. 60). Common forms of this are to state a preference or the use of an interrogative form in order to convey a request. For example:

I:4.5   *Teacher:*   I'd like to take in your exercise books.

I:4.6   *Diner:*   Can you bring me the wine list?

These are both instances of how speakers frequently accomplish an indirect speech act by stating or questioning one of the felicity conditions (Gordon and Lakoff,

1971). In I:4.5 the teacher explicitly states his wish to receive the books, i.e. that he meets the 'sincerity' condition (Levinson, 1983); in I:4.6 the diner questions the *ability* of the waiter to provide the list i.e. s/he questions one of the preparatory conditions (*ibid.*). Note that I:4.6 is only conventionally interrogative, in that the waiter does not have the option of treating the question as a request for information rather than for action. I shall give particular attention to these matters as features of mathematics talk in Chapter 8, in the context of a Theory of Politeness.

Traditionally, three major language functions are identified – statement, question and command – having typical realizations in declarative, interrogative and imperative verb forms. Indirect speech acts break down these canonical correspondences between language function and form. Thus I:4.5 and I:4.6 achieve commands through declarative and interrogative forms respectively. Sinclair and Coulthard observe that:

> Modal verbs play a large part in producing the lack of direct correlation between the three grammatical forms and functions. (1975, p. 11)

Indirect speech acts are one of the means whereby a speaker may convey propositional attitude whilst at the same time making a declarative utterance. Consider an imagined scene from the mathematics classroom. The teacher singles out one child, and asks, 'How many lines of symmetry does a rectangle have?', The child might simply answer, 'Two.' But it is possible to imagine circumstances in which the child may wish to convey that its answer is tentative, controversial, questionable or whatever. Some possible formulations include:

I think it's two.

Two, maybe.

Basically, it's two.

I'd say it was two.

My Dad said it was two.

In each of these cases, the illocutionary force of the utterance is hedged so as to convey doubt whether 'two' is the correct answer, and to withhold full commitment to it. There is associated with each of them a 'hedged performative' (in some cases, tacit), so that what in a more confident speaker would be a statement of fact ('There are two') becomes an action signifying uncertainty. I shall show how hedged performatives are used in mathematics talk to express modality and convey speaker attitude to mathematical propositions.

Conversational Implicature

I come now to the notion of conversational implicature, due to the philosopher Paul Grice (1975, 1989), who proposed that ordinary conversation is posited on a

Co-operative Principle (CP), whose meaning is embodied in four sub-principles or 'maxims' of conversation. These maxims specify what participants need to do in order to converse rationally and co-operatively. The requirements are, essentially, maxims of:

- *Quality*.[4] Let your contribution be truthful: do not say what you believe to be false.
- *Quantity*. Let your contribution be as informative as is required (for the current purposes), and not more informative than is required.
- *Manner*. Let your contribution be clearly expressed, e.g. be brief, orderly, unambiguous.
- *Relevance*. Let your contribution be relevant to the matter in hand.

Now it is evidently *not* the case that all participants in all conversations observe all these four maxims in all contributions. More often than not, this has nothing to do with an intention to lie or mislead (in which case the participation might not be deemed co-operative). Conversation can and does include utterances – such as refusals, disagreements and abuses – which would not be regarded as 'co-operative' in the ordinary meaning of the word. In what sense, then, does Grice use the word, given that participants will be expected (1989, p. 26) to observe the CP? Grice himself does not attempt much of a gloss on his intended meaning:

> We might then formulate a rough general principle which participants will be expected (ceteris paribus) to observe, namely: Make your conversational contribution such as is required, at the stage at which it occurs, by the accepted purpose or direction of the talk exchange in which you are engaged. One might call this the Cooperative Principle. (*ibid.*)

One can proceed from this start in two ways. Mey adopts a position which holds that a contribution may fail to be co-operative. At least, that is the clear message of his telling and analysis of an anecdote in which a family 'friend' is abominably rude to his 6-year-old daughter. Irritated when Sarah loses her ball in his book-lined study, he flaunts his erudition, to the bewilderment of the child, by directing her to 'look behind Volume 6 of Dosteyevski's collected works'. My human sympathy is with the father, who tersely comments (Mey, 1993, p. 67):

> the adult interlocutor failed to observe the principal demand set up by Grice in the CP: namely, to cooperate with your conversational partner.

On the face of things, the Dosteyevski utterance violates the maxims of Manner and of Quantity. There is, however, an alternative reading of the incident. Being co-operative is not the same thing as being pleasant. Grice is sparing in his commentary on the meaning of 'co-operative', but one could suggest that the essence of co-operation is an intention to assist other participants in their pragmatic efforts to make human sense of the spoken interaction. Thus, in being angry or nasty, or in making a complaint, we may fail to be 'nice', but we certainly assist the communication of our meanings, feelings and attitudes. In fact, Mey may be mistaken in his assumption that the friend's remark (which Mey takes as unco-operative) is addressed to the *child*. Surely the intended audience was in fact the father, and the intended implicature something like 'I'm irritated because your daughter is playing in my study', or perhaps 'Please remove your daughter from my study'.

How, then, do hearers interpret speech which is supposedly co-operative yet which is, for example, superficially untrue or irrelevant? Viewed from the other side of the coin, how do speakers successfully violate the maxims in order to communicate fine nuances of meaning, to enable the hearer to read 'between the lines', as it were?

The genius of Grice's theory is the following recognition. Whilst speakers do not always observe the maxims at the surface level, nevertheless hearers interpret the contributions of other participants in conversation as if they were intended to observe the maxims at some level of meaning other than that contained in the semantic content of the utterance. This has proved to be a robust theory, concise yet surprisingly complete, with a wide field of application, finding resonance with common sense and experience. For Grice describes and explains what we all know – that 'communication involves the publication and recognition of intentions' (Sperber and Wilson, 1986a, p. 24).

Such a view of communication underpins a means of pragmatic inference identified by Grice, which he named 'conversational implicature'. On many occasions I shall show this means of inference in action in mathematical conversation. To clarify the meaning and process of implicature, consider the following exchange:

I:4.7 *Teacher:* Why haven't you brought your calculator to my lesson?

I:4.8 *Pupil:* My brother has a maths exam today.

The pupil's reply, taken literally, is irrelevant to the teacher's question. Indeed, the pupil appears to be flouting the maxim of Relevance. We could interpret the pupil's contribution as simply non-co-operative, failing to address the teacher's enquiry about the missing calculator. In practice, we interpret the exchange as co-operative at some level, albeit not a superficial one; we infer, from the ostensibly irrelevant I:4.8, that:

I:4.9 My brother has my calculator, because he needs it for his exam.

The inference is an example of a (conversational) *implicature*, and we say that I:4.8 *implicates* the conclusion I:4.9. An implicature is rather like a hint. Thus the human tendency is to accept Grice's theory as an accurate insight, one that exposes, codifies and resonates with that which we 'knew' but had not yet isolated.

It is interesting to take the analysis one stage further, to consider the mechanics of pragmatic inference, i.e. how implicatures may be 'calculated'. Sperber and

Wilson (1986a) comment that the hearer brings his or her 'cognitive environment' to bear on the situation, for the purpose of interpretation. This environment consists of a (very large) set of facts which are manifest (*ibid.*) to individuals by knowledge or by assumption (including, for example, the materials usually needed by candidates in a maths exam). It is on the basis of selection from this environment, and not from the speaker's utterance in isolation (if at all), that inferences are to be made. Pragmatic implicature is very different from logical implication, in that the inferred conclusion – I:4.9 in the example – cannot be obtained solely from what has actually been uttered by the application of a syllogism, or other process of logical deduction.

On the contrary, an implicature may be semantically contradictory to the utterance that gives rise to it. Grice (1989) shows how a number of figures of speech can be viewed in terms of speakers *flouting* or *exploiting* one or more of the maxims for the purpose of making an implicature. Consider irony, for example. Suppose that I am hoovering the carpet, when my son comes in with muddy shoes. I might say, 'Thanks, that's a big help!' Since my son is aware that he is in fact frustrating my domestic purposes, it is clear to him that I cannot believe what I am saying, i.e. I am in breach of the maxim of Quality. Assuming that my intention is nonetheless co-operative, he casts around for a *related* proposition that I might be wishing to convey. The negation, that he is hindering rather than helping, is one such possibility, and it is Relevant to the situation. My son thus infers my displeasure as an implicature. Crucially, if there were no fundamental assumption of co-operation, ironies would be pointless utterances, and no inferences could be drawn from them.

Gazdar (1979, pp. 49–52) accounts for Quantity implicatures in particular with the following formalization. Let a, b, c and d be sentences such that

- d entails a;
- b is the negation of d;
- c is the conjunction of a and d (usually including d as a suspension clause);
- b is an implicature of a.

Among several examples, Gazdar includes:

a. I believe he's ill.
b. I don't know that he's ill.
c. I believe, in fact I know, that he's ill.
d. I know that he's ill.

The inclusion of the suspension clause (affirming d) in c has the effect of 'cancelling' the implicature b. The relation between a and b cannot be one of entailment, since if it were (by transitivity) then d would entail its negation, b. But the a sentence is weaker than the d sentence, which, if it could be asserted, would be more informative than a. Anyone uttering a (whilst being co-operative) must therefore be implicating the negation of d, which is b.[5]

Similarly, the hedged promise:

I:4.10   Maybe I'll come and visit you next week

flouts the maxim of Manner, and implicates the related proposition

I:4.11   I may fail to come to see you next week

because, if I were firm in my desire and intent to come, I would have made an unambiguous promise without hedging the assurance. Grice's theory of implicature thus accounts for the effectiveness of hedges in conveying uncertainty, lack of commitment.

A general formulation for working out, or 'calculating', an implicature, can be summarized as follows (Levinson, 1983, p. 113). Suppose S has said p. We expect S to be observing the CP and the maxims of conversation. In that case, given that S said p, he could intend me to understand that q. Moreover, S has done nothing to prevent me from thinking that q is the case, and he knows that my thinking q will enable me to preserve the expectation that he is co-operating. Therefore S intends me to think q; which is implicated by his saying p.

A useful way of looking at the Gricean interpretive framework would be to say that either speakers are overtly co-operative because they observe the maxims, or else they are covertly co-operative by setting up implicatures by means of maxim violations. In this technical sense, though not in the everyday sense, one can argue that every contribution to conversation is 'co-operative'. This is convenient in that Grice's maxims then become absolutes rather than mere ideals. Whenever possible, I shall approach the CP and the maxims with the expectation that speakers intend their interlocutors to make sense of what they say, even though they will not always (superficially) simply *say* what they mean.

Politeness Theory

In Chapter 1, I introduced a discussion between Judith and 14-year-old Allan.[6] In the course of a mathematical investigation, Judith asks:

IRG5:51   *Judith:*   Right. Can you make any predictions before you start?

The indirect form of Judith's request for a prediction (questioning a felicity condition, i.e. the boy's ability to provide it) is very characteristic of many of the teachers whom I studied through transcripts. In fact the boy made a prediction, but the vagueness of his answer suggests that it was far from secure:

54   *Allan:*   The maximum will probably be, er, the least 'll probably be 'bout fifteen.

A particular quality of insight into such indirectness and vagueness in classroom mathematics talk is provided by a sociolinguistic theory – of 'politeness' – due to Penelope Brown and Stephen Levinson (Goody, 1978; Brown and Levinson, 1987).

In essence, politeness theory is constructed to account for some indirect features of conversation; it claims that speakers avoid threats to the 'face' of those they address by various forms of vagueness, and thereby implicate their meanings rather than assert them directly.

Politeness theory is based on the notion that participants are endowed with two properties – rationality and 'face'. 'Face' (Goffman, 1967) consists of a public self-image, with two 'wants':

- positive face – a desire to be appreciated and valued by others; desire for approval;
- negative face – concern for certain personal rights and freedoms, such as autonomy to choose actions, claims on territory, and so on; desire to be unimpeded.

The Model Person (MP) of the theory not only has these wants her/himself, but recognizes that others have them too; moreover, s/he recognizes that the satisfaction of her/his own face wants is, in part, achieved by the acknowledgment of those of others. Indeed, the nature of positive face wants is such that they can be satisfied only by the attitudes of others.

Now some acts ('face-threatening acts', or FTAs) intrinsically threaten face. Orders and requests, for example, threaten negative face, whereas criticism and disagreement threaten positive face. The MP therefore must avoid such acts altogether (which may be impossible for a host of reasons, including concern for her/his own face) or find ways of performing them whilst mitigating their FTA effect, i.e. making them less of a threat.

Imagine, for example, that someone says something that MP believes to be factually incorrect. MP would like to correct them. Such an act would threaten the first speaker's positive face – the esteem in which s/he is held as a purveyor of knowledge. Or suppose that MP would like someone to open the window, but is aware of the threat to the other's negative face. Brown and Levinson identify a taxonomy of strategies available to MP in such circumstances.

1   *Don't do* the FTA – simply agree or keep quiet.
2   *Do* the FTA: in which case there is a further choice of strategy:
    2.1   Go *off record* – don't do the FTA directly, but implicate it, e.g. 'Don't you think it's hot in here?' (indirect request to open the window).
    2.2   Go *on record:* either
       2.2.1   *'baldly'* – essentially making no attempt to respect face; or
       2.2.2   with *redressive* action: having regard either for the other's
          2.2.2.1   *positive* face ('You're the expert in these matters, but I thought that . . .'; or
          2.2.2.2   *negative* face ('I'm sorry to trouble you, but would you mind . . .')

Figure 2   Politeness strategies (after Brown and Levinson, 1987)

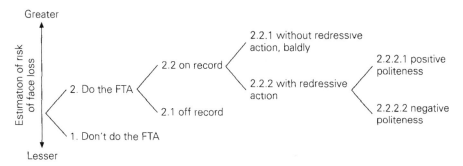

These strategies are summarized in Figure 2 (Brown and Levinson, 1987, p. 69).

Going on record baldly suggests greater concern for one's own face than for that of one's audience. I associate it with being 'assertive'; it sounds like bullying, but it may be the only way that someone in a position of weakness can avoid being ignored or bullied themselves.

Redressive action is very commonplace when FTAs are in prospect; such action is a way of indicating that no face threat is intended. The strategies in the lower part of the diagram offer the least such threat, those in the upper part the greatest.

It is important to note that whereas an utterance (request, criticism, etc.) most obviously stands to threaten the face of the addressee (H), it may in fact threaten the face of the speaker (S). For example, making an excuse or accepting an offer may offend S's negative face (s/he feels obliged to justify, s/he is in debt to someone). A confession or emotional outburst offends S's positive face (s/he is seen as less worthy, less in control).

Sociological dimensions of 'face'

Brown and Levinson (1987, pp. 74–84) – building on work of Brown and Gilman (1970) on the social connotations of second person pronoun use – isolate three factors in the assessment of the seriousness of an FTA. These factors are:

- the 'social distance' (D) of S and H;
- the relative 'power' (P) of S and H;
- the absolute ranking (R) of impositions in the culture.

D, P and R reflect the extent to which the protagonists (S and H) *perceive* these factors as mutual knowledge, as opposed to some rating of actual power. A value for each of these three factors is assessed (notionally, on a scale of 1 to *n*), against the following descriptions.

D is a symmetrical function D(S, H) which measures social proximity. Notions of class and educational achievement, for example, may be factors in the assessment of D. More fundamentally, D depends on frequency of interaction and mutual exchange of goods (material and non-material). A high value of D, reflecting close social proximity, is generally associated with mutual concern for the other's positive face – such as concern for the self-esteem of a colleague.

P is an asymmetrical function P(H, S) which measures the relative power of H over S, in particular the degree to which H can impose his/her own wants at the expense of S's. The sources of P are either material (such as physical force) or metaphysical (by virtue of 'powers' ascribed to H). Usually, both will be relevant. A high value of P will generally be associated with deference – such as a pupil might be expected to show to a teacher, or a teacher to the head teacher.

R is a culturally and situationally determined constant, R_x, indicating the ranking of the imposition of the FTA x. Thus an employee's request for 'time off' on account of a hospital appointment might (in appropriate circumstances) be ranked lower than a similar request to attend a child's school play.

The dependence of R on context is clear; Brown and Levinson also argue for the context-dependence of D and P. For example, a high-rated social distance (in the context of the home town) between a parent and his child's teacher might significantly lessen if they found themselves in the same Mediterranean holiday hotel. In addition, Brown and Levinson argue for the independence of the three social factors, and thereby justify the simple summation of D, P and R in a given situation, in the assessment of an FTA. The sketch that I have given will be sufficient for application to the analysis of classroom mathematics talk.

Indirectness

Brown and Levinson identify and catalogue a number of linguistic strategies associated with the face-respecting options shown in the diagram above. These include the use of indirect speech acts, including quasi-interrogative forms such as:

I:4.12   You couldn't just pop out and get me a newspaper, could you?

which redress the threat to the addressee's negative face, their autonomy, respecting their right to refuse. However, given high values of D and P, for example, there may be mutual recognition that refusal is not a real option. The following example is from one of a number of interviews with 10 and 11-year-old children, which I shall examine in detail in Chapter 6.

T3:17   *Tim:*   OK, now let's think about two numbers that add up to twenty. Would you like to start off, Caroline?

Both Caroline and I know that this is an instruction, that the indirectness (marked here by the avoidance of the imperative) is *conventional*. Caroline knows that she has no option but to 'start off'. I am, nevertheless, sincere in my wish to be seen by these young students to be gentle, considerate and non-threatening. I shall draw attention to the prevalence of this kind of indirectness in Chapter 8.

Conversation Analysis

Broadly speaking, the pragmatic theories that I draw on derive from the philosophical tradition of language analysis, which is mainly associated with sentence structure and the logic of meaning. These latter semantic interests have most recently extended into what Mey (1993) calls 'micropragmatics', by which he means those aspects of language use which are analysed by reference to individual sentences or utterances, or at most pairs or short sequences of such units of text. A number of philosophical contributions to pragmatics at this level have been outlined in this chapter, notably the notions of speech act and implicature.

One characteristic of the philosophical approach to language is the status of invented (or intuitive) sentences as data, an approach somewhat alien to the late twentieth-century education research tradition in which sociology and anthropology are so influential – though not, perhaps, to a pure mathematician, whose only 'data' are self-generated. For the philosopher, introspection and appeal to intuition are valid elements of method and sources of data. The following invented exchange is quoted from Grice's (1989, p. 32) exposition of conversational implicature:

> *A:* Smith doesn't seem to have a girlfriend these days.
> *B:* He has been paying a lot of visits to New York lately.

No source (such as a corpus of conversation) is cited, and it is assumed that none exists, other than Grice himself. The argument (that B implicates that Smith has, or may have, a girlfriend in new York) is not dependent on the claim that this is a fragment of a 'real' conversation. Similarly, Brown and Levinson's account of off-record FTAs proceeds (1987, p. 69):

> So, for instance, if I say, 'Damn, I'm out of cash. I forgot to go to the bank today,'
> I may be intending to get you to lend me some cash, but I cannot be held to have
> committed myself to that intent [. . .]

Is the conclusion ('I cannot be held,' etc.) any weaker because no tape recording exists of a spontaneous utterance 'Damn, I'm out of cash,' etc? Gazdar (1979, p. 11) is explicit:

> I shall assume . . . that invented strings and certain intuitive judgements about
> them constitute legitimate data for linguistic research.

Philosophers such as Austin and Grice have usually based their arguments on consideration of such 'invented strings'. Nevertheless, such an approach has some limitations. The most obvious is that the audience for the argument must accept the plausibility of the invented text; in effect, to agree that the proposed sentence or sentences might well occur in some actual discourse.

Harvey Sacks laid the foundation for an overtly empirical approach to the analysis of discourse in a series of lectures from 1964 to 1972; Harold Garfinkel coined the term 'ethnomethodology' with reference to inductive study of naturally occurring speech data. With this perspective, Sacks developed the approach to discourse termed conversation analysis (CA). The term 'talk-in-interaction' (Schegloff, 1987) has gained acceptance to indicate the scope of CA. Some of the most notable contributions of CA have been achieved in particular institutional settings, such as law courts (Atkinson and Drew, 1979) and medical interviews (Frankel, 1990).

CA came into being as a response to some perceived inadequacies of Searle's speech act theory – arguably one of the philosophical frameworks *most* sensitive to social organization, and offering great promise in the illumination of particular institutional practices. The work of Labov and Fanshel (1977) attempted to apply speech act analysis in a study of psychotherapeutic interviews. Such interaction is typically indirect, and meanings emerge from sentence or utterance. One essential shortcoming of the speech act approach to the analysis of discourse exposed in this study is the tendency to look for the acts performed by small units of speech. As Drew and Heritage comment:

> There can, by now, be no serious doubt that sentences and utterances are designed and shaped to occur in particular sequential and social contexts and that their sense as actions derives, at least in part, from such contexts. (Drew and Heritage, 1992, p. 12)

A defining feature of CA is the approach to conversation in terms of extended sequences of utterances. Such sequences themselves contribute to a rich view of 'context', which construes utterances (and the social 'acts' they perform) as doubly contextual (Heritage, 1984). First, such utterances and actions are *context-shaped*, in that their production takes place both in a local configuration of speech (the *co-text*) and action which precedes them, and also in a wider spatial, temporal and inter-personal environment which contains that configuration. Second, utterances and actions are *context-renewing*, forming the immediate context of some next action in a sequence. In this way, the CA analyst takes a dynamic approach to context, in contrast to the 'bucket' theory, which prescribes and proposes a fixed framework of context to account for certain features of a given discourse.

A speech act-based approach to discourse which takes account of the sequential organization of action had in fact been developed by Sinclair and Coulthard and their collaborators in Birmingham. In their study of classroom discourse (1975) a model of interaction was developed in terms of sets of acts, moves, exchanges and transactions. One such regularity associated with teacher questioning is the three-part Initiation–Response–Feedback (I–R–F) cycle, the prevalence and relevance of which to mathematics classrooms has been widely acknowledged since Sinclair and Coulthard first drew attention to it. In one example (due to Pimm, 1987, p. 27), a teacher Initiates a cycle by asking his class how a route map on the blackboard might be communicated to someone in the next room. A pupil gives a one-word Response, 'Co-ordinates.' The teacher follows up with evaluative Feedback, 'That would be a very good way of doing it,' and immediately Initiates the next cycle with another question, 'What do you mean by co-ordinates?'

Such attempts to specify discourse text in terms of such 'rules', to develop a kind of grammar of interaction, are viewed with suspicion by CA proponents such as Drew and Heritage:

> In their preoccupation with the rules for discursive action within a context, the Birmingham group tended to ignore the task of analyzing how mutual understandings are achieved by the participants [. . .] This engendered a related failure to specify in their model how participants show their orientations to the particular institutional context in which they are interacting. For example [. . .] their analysis failed to disclose the ways in which successive elements of the I–R–F sequence *constitute* its instructional character. (1992, pp. 14–15)

Drew and Heritage appear to discount the value of the Birmingham research in *sensitizing* the transcript analyst to the structure of the text s/he is examining, and the opportunity it presents for expert theorizing about the relevance of the 'rules'. For example, Pimm (1987) makes it his business *precisely* to illuminate, with reference to particular fragments of transcript data, how the I–R–F sequence exposes 'how participants show their orientations to the particular institutional context in which they are interacting' (e.g. the teacher's desire for control, the pupil's sense of 'exam' questions in the classroom), and conversely how the orientations can be seen to give rise to the I–R–F sequence. Consideration of these same orientations might suggest that the I–R–F sequence is not a regularity to be found in, for example, mundane dinner-party conversation.

CA rejects the premature construction of theories and any consequent deductive inferences from them. The CA approach is fundamentally *inductive*, although this is by no means a distinguishing feature of CA. In keeping with the interpretive research paradigm, there is constant dynamic interplay between data, theory and analysis (Hopper, 1989, p. 52). Data consist of tape recordings and very detailed transcriptions of those recordings, coding pauses, intonation, sound stretches, interruptions and overlaps, dysfluencies and so on. This renders fluent reading of such transcripts difficult, but that is not the transcriber's prime objective.

A piece of CA: Adjacency Pairs and Preference Organization

A major object of inductive attention in CA is the organization of sequences of utterances, and in particular the sequential partition of conversation into 'turns' (in the sense that participants 'wait their turn'). This has led to the creation of a literature on the management of conversational turn taking, and the related notions of directionality (projection from one participant to another) and inter-subjectivity (so that next actions display a particular understanding of prior contributions).

These notions came under consideration in discussions of the concept of adjacency pairs (Schegloff and Sacks, 1973) and preference organization. The notion of an adjacency pair arises in consideration of paired utterances. A pair is initiated by one speaker with a 'first part':

I:4.13   *A:*   Can you come over for coffee tomorrow?

Table 1   Adjacency pairs: correlation between parts

First part	Request	Offer/invite	Assessment	Question	Blame
Second part					
Preferred	accept	accept	agree	expected answer	denial
Dispreferred	refuse	refuse	disagree	unexpected	admission

which is followed by the second part:

> I:4.14   *B:*   Yes, thank you, I'd love to

or perhaps:

> I:4.15   *B:*   Well, er, let me see, I have to go shopping in the morning, and my brother said he might come over some time, so it's a bit tricky, really.

Some types of second parts are routinely 'dispreferred' in response to first parts such as questions, offers, requests. I:4.15 is a more complex, hesitant response than I:4.14, which is a 'preferred' second part, in this case acceptance of the invitation. The complexity of the dispreferred second part is an observed linguistic (CA) regularity. The more elaborate structure of I:4.15 marks it as 'dispreferred' (non-acceptance).

Levinson (1983, p. 336) lists a correlation between different types of first parts and preferred/dispreferred second parts (Table 1). Characteristics of dispreferred seconds include (Atkinson and Drew, 1979):

- *delays:* such as pauses before delivery;
- use of *prefaces:* particles such as 'well', token agreements (I'd like to, but . . .);
- *accounts:* carefully formulated, over-elaborate explanations for the dispreferred act.

Here, for example, a teacher (Hazel) asks a 10-year-old pupil (Faye) a question which presupposes a certain belief. Faye's reply is delayed and prefaced, marking it as a dispreferred second pair part:

> IRG3:33   *Hazel:*   . . . why do you think that for certain?
>   34   *Faye:*   Because . . . well, I don't know for certain but I think . . . 'cos the numbers that we've done are quite close to the first . . .

The second part disagrees with an assessment presupposed by the question. It is also an unexpected answer to a question.

The complexity of my intuitive example I:4.15 above (consistent with the characteristics of dispreferredness) can alternatively be interpreted in terms of (1) violation of the maxim of Quantity or (2) redress of a threat to A's positive face. Indeed, the characteristics of dispreferred seconds listed above are all consistent with the avoidance of FTAs; much of politeness theory could be reconstructed within CA. Levinson (1983, pp. 345–64) has devised an ingenious CA reinterpretation of the notion of indirect speech act as a particular type of 'pre-sequence', on the basis of this framework of turn taking and preference.

Overview: Approaches to Discourse

Levinson (1983) presents CA as an approach to discourse which is in opposition to less empirically based approaches. His somewhat adversarial approach has since been taken up by supporters on both 'sides' (see, for example, Hopper, 1989, p. 60; Mey, 1993, pp. 48–49). This is unfortunate, because each approach offers different kinds of insights into interactive talk, and the wise analyst would do well not to be exclusive when interpreting transcripts.

Deborah Schiffrin (1994) sets out to reconcile many of the differences between alternative traditions in discourse analysis. She identifies six such approaches, including those associated with speech act theory, with the Gricean co-operative framework, with interactional sociolinguistics (embracing Politeness Theory) and with ethnomethodology. Schiffrin illustrates the different ways that these approaches illuminate sample texts. She argues that

> All [these approaches] attempt to answer some of the same questions, e.g. How do we organise language into units that are larger than sentences? How do we use language to convey information about the world, ourselves and our social relationships? (p. viii)

Schiffrin concludes that:

> [. . .] all the approaches to discourse view language as social interaction, and all are compatible with a functionalist rather than a formalist paradigm. (p. 415)

One of six unifying principles identified by Schiffrin (p. 416) asserts that analysis of discourse is empirical; analyses are predictive, they produce hypotheses. This accurately characterizes my own approach to transcripts of mathematics talk.

Elsewhere, Brown and Yule (1983, p. 22) similarly claim that:

> The discourse analyst, with his 'ordinary language data' [. . .] may wish to discuss, not 'rules', but **regularities**, simply because his data constantly exemplifies non-categorical phenomena. (Author's emphasis)

This interest in regularities is indicative of my own approach to the analysis of the transcripts of mathematics talk. But, for me, those regularities must have some inferential significance; I cannot be content merely to describe them. I *do* want to try to get 'behind conversation', to make inferences and conjectures about 'what is really going on' (Levinson, 1983, p. 287) – transactionally and interactionally – in mathematics talk.

My analytical approach is eclectic. My study of transcripts has been highly inductive. I come to the transcripts without particular *linguistic* expectations. I describe what I find there, and then I set about studying that regularity, that phenomenon.

Summary

In this chapter I have laid out the linguistic elements of the analytical framework that I shall apply to the analysis of many mathematical conversations. Speech is perceived as performing actions of various kinds, the effects of which on others may be indirect. Similarly, the intended meanings of utterances may be superficially obscure. Grice's theory of implicature asserts that co-operation (in the sense I have described) is nevertheless assured in conversation, and that this principle is central to the very possibility of pragmatic interpretation of various kinds of vague and indirect contributions to conversation.

Whereas considerations of politeness can be argued to account for many vague and indirect features of conversation, these same features can be viewed (from a CA perspective) as an empirical regularity of interactive talk.

So much for the framework. At last we come to some data. In the next chapter I introduce some conversations with children, and demonstrate the crucial communicative and meaning-making function of deictic language elements of mathematical discourse.

Notes

1 There is a danger of over-simplifying Glasersfeld's account for the sake of brevity; he describes the co-ordination of 'sound-image' and 're-presentation of experience'. This co-ordination is also encapsulated in the Saussurean notion of 'sign', consisting of both signifier and signified.

2 Vestiges of the singular 'thou' survive in speech in northern England, and non-standard English includes plural forms of 'you' in various dialects, e.g. 'y' all' (United States) and 'you'se' (Ireland).

3 Anaphora and deixis are not necessarily mutually exclusive: Levinson (1983, p. 67) gives the example 'I was born in London and have lived there ever since', in which 'there' is co-referential with 'London' (anaphora) but also locates (deixis) the utterance outside London. The deictic effect is achieved by the choice of 'there' rather than 'here'.

4 I adopt the convention of using capital letters for Quality, etc., to mark the maxims. The esteem in which Grice's theory of implicature is held is indicated, in my view, by a number of attempts to revise and 'improve' it by proposing more economical alternatives. In particular, Horn (1984) has just two principles: 'Q' comprises Quality with the 'not too little' component of Quantity, and 'R' is the 'not too much' part of Quantity

with Manner and Relevance. Sperber and Wilson (1986a) are more radical, proposing a super-maxim of Relevance, along with a concept of 'mutual manifestness', to subsume them all. I judge that what Grice's original formulation lacks in economy is compensated for by its transparency, and prefer to adhere to it.

5 Whereas Gazdar's argument draws on the maxim of Quality alone, it does also seem to require reference to Quality. (The speaker can't assert the more informative d sentence because its propositional content exceeds that which s/he knows to be true.)

6 A full discussion of Judith's conversation with Allan follows in Chapter 8.

5   Pointing with Pronouns

> What is implied in the proper use of pronouns? Do children recognise them early
> and integrate them in their own speech with ease and total comprehension?
> (Gattegno, 1981, p. 5)

In this chapter, I shall show how the contribution of context to the interpretation
of vague utterances enhances the ability of speakers (particularly novice speakers
of mathematics) to refer to aspects of their own mathematical thinking, and thus
assists their communication of such ideas. The argument centres on the use of
pronouns in mathematics talk, and is based on two generative case studies.

The Informants

My transcript data collection began with extended, weakly framed mathematical
conversations with two children, Susie and Simon. My technique in contingent
questioning was substantially trialled and developed in a series of interviews with
Susie. I then undertook a somewhat smaller case study applying the same weakly
framed interview method; the subject of these three interviews, which took place
over a Christmas vacation, was Simon, aged $12^{3}/_{4}$ at the time. Like Susie, he is a
quick thinker, and he articulates his thinking well. Paradoxically perhaps, being in
close family relationship with me, Simon probably sees me as an 'authority' figure
in mathematical matters more than Susie does; he is aware of my background and
what I do at 'work'.

Yet Simon plays only a supporting role in this story, whilst Susie has the lead.
It was the conversations with Susie that first breathed life into the research, and
affirmed my belief in the value of close study of individual children. As Stephen
Brown says:

> One incident with one child, seen in all its richness, frequently has more to convey
> to us than a thousand replications of an experiment conducted with hundreds of
> children. (1981, p. 11)

Susie was 9 years old at the time we began our extended mathematical conversations. We talked together for over three hours in total, initially on four occasions
over a period of five weeks (transcripts S1–4) and again some months later (S5).[1]

I had arranged to spent most of one spring term with her class at her school.
Her teacher favours children working and learning together, but Susie was a child
who did not seem to thrive in a co-operative group situation; she rarely seemed to

be impressed by the ideas of her peers. Conversely, her proposals were usually ignored by them – partly, I think, because her insights were frequently inaccessible, or were elaborated at a length beyond the attention span of her audience.

However, I was frequently fascinated by her contributions to teacher-managed class discussions when she showed that she was articulate and willing to expose her thinking to external scrutiny: for example, in a debate to determine the cost of one ruler, ten of which cost £3.50, Susie volunteered, 'It's 35p, 'cos you cross off a nought.' This was immediately followed by other estimates and proposals which suggested that very few of the children had listened to Susie's contribution, or had regarded it as being especially significant. My own instinctive reaction was not that she had chosen an appropriate operation, but that she was 'merely' rehearsing a rule she had learned somewhere to execute the calculation. Perhaps with the same thought in mind, her teacher returned to Susie, and invited her to say more to the class about her method. 'You cross off a nought,' repeated Susie, continuing: 'If you have ten, and you take away nine ones, you have just the one left . . . It's because you take away a ninth . . . no, nine-tenths. So there's one-tenth left.' I was, and am, fascinated by the fact that Susie 'explains' division by ten by talking about subtracting nine-tenths i.e. take away nine-tenths of ten (she seems to be saying) and you're left with one, just like crossing off the nought. Elsewhere (Rowland, 1997a), I have described and analysed Susie's idiosyncratic approach to a particular class of division problems.

On this and on other occasions, Susie gave evidence of confidence, efficiency, and an unusual self-monitoring capability in her mathematical thinking, As it happens, her reading and writing of English were at that time significantly behind her mathematical, scientific, and indeed her artistic attainments.

She had another quality which was invaluable for my research purposes: a quiet but determined intellectual independence (not to put too fine a point on it, stubbornness), coupled with a direct kind of honesty which can manifest itself as rudeness, as measured by conventional social norms. Consequently, I never felt, in our conversations, that Susie was saying what she thought might please me in preference to what she believed. She frequently interrupted me when it suited her to do so. I initiated our conversations, but I didn't feel that I controlled them. To this extent they were weakly framed as well as 'contingent': Susie seemed very happy to think on her feet, and I was compelled to do the same. On inspection of the text, I was confronted with a significant linguistic phenomenon – Susie's use of certain pronouns.

The Exclusive 'We'

In Chapter 4 of *Speaking Mathematically*, David Pimm discusses the use of the pronoun 'we' in adult social practice, in particular in mathematics pedagogy. The following dialogue is from a classroom excerpt (Pimm, 1987, p. 65) involving a teacher and a 10-year-old pupil. The problem under discussion is 26 – 17.

> *Teacher:* What column's that? The tens column. Right. And what do **we** do there?
> *Pupil:* **We** cross that one out . . . and then **we** put one there.
> *Teacher:* **We** take a . . .
> *Pupil:* Er . . . er . . . er . . .
> *Teacher:* **We** take a . . . What do **we** take from the tens column? **We** take a ten, don't **we**?

It is a ritual intonement of a procedure (in this case, for subtraction) which has been imposed on the audience, the child. This particular algorithm (decomposition) seems to be a particularly rich source of teacher 'we's. Hilary Shuard transcribed a similar conversation (given in full in Shuard, 1986) around the 'sum' 42 − 25, which includes:

> *Teacher:* [. . .] Why can't you do it?
> *Pupil:* Two ones.
> *Teacher:* You've only got two ones, haven't you? You haven't got enough. Do you remember, when **we** went through these sums last week, what **we** said you had to do, if you hadn't got enough?
> *Pupil:* . . . er . . .
> *Teacher:* What did **we** say you had to do?

It is improbable that the child is included in the 'we' in phrases like 'what we said you had to do'. The phrase could be intended to imply 'What I said in your presence'. Mühlhäusler and Harré (1990, p. 129) propose that when academics use 'we' in exposition, it is in order to draw the listener into complicity. They point out that the addressee is thereby trapped in tacit agreement, and so prevented from voicing opposition by the special (if ephemeral) relationship that has been artificially forged between expositor and audience.[2] I believe that the teacher's 'we' has much of this quality. Pimm (1987, pp. 69–70) suggests that the teacher is often associating herself with some other (unnamed) person or persons. He argues that the teacher, by using the plural pronominal form, is sometimes appealing to an unnamed 'expert' community to add authority to the imposition of a certain kind of classroom practice. Like the editorial 'we' (Wales, 1980, p. 27), the effect is to associate the speaker with a select and powerful group, from which the audience is clearly excluded. The result is to discourage and devalue any sense the child may make of the situation, and to urge acquisition of the 'proper way' of doing such 'sums'. Such appeals to the support and authority of unspecified others is not, of course, peculiar to mathematics. Wills observes that:

> 'We' seems to have the greatest *imprecision of referent* of all English pronouns, and therefore is the most exploited for strategic ends. (Wills, 1977, p. 279, emphasis added)

I now turn to consider some ways in which Susie used two particular pronouns in our mathematical conversations.

'It'

On examining the transcripts of my conversations with Susie, I became aware of the frequent appearance of the pronoun 'it' in our dialogue. By way of illustration, here is an extract from our third session.

S3:68	*Tim:*	What about this one you did; 260 divided by ten is what?
69	*Susie:*	Twenty-six. [Tim writes $260 \div 10 = 26$]
70	*Tim:*	Right. And what's twenty-six times ten?
71	*Susie:*	Twenty-six times ten . . . twenty-six lots of ten . . . ten lots of twenty-six . . . oh, **it's** with forty **it** doesn't work. With forty I don't think **it** . . . except with ones and tens and . . . ones and tens. **It** wouldn't . . . and twenties, sometimes twenties, . . . em, sometimes thirties, sometimes forties, sometimes fifties, sometimes sixties, sometimes seventies, eighties, nineties, . . .
72	*Tim:*	You mean . . .
73	*Susie:*	If you do ten . . . and I think **it** would be ten if **it**, . . . I don't, I'm not sure . . . suppose you had 266 . . . I'm not sure about this, I'll just find out.
74	*Tim:*	Yes, you experiment and find out [Susie writes $266 \div 10 = 26.6$]
75	*Susie:*	That's twenty-six point six. And twenty-six point six lots of ten . . . so you'd in a way put a nought on the end, but you'd end up like that. [She has written $26.6 \times 10 = 266$]
76	*Tim:*	When you say . . . [interrupted]
77	*Susie:*	So any tens with any other, with any number, **it** would end up like that.
78	*Tim:*	With tens . . . [interrupted]
79	*Susie:*	[*forte*] and the same with ones, but not with something like sevens, or whatever. And sometimes with twenties and thirties and forties and fifties and sixties and so on.
80	*Tim:*	What's another number like seven that you think **it** wouldn't work with?
81	*Susie:*	**It** wouldn't work with . . . [Writes $30 \div 7 = $]
82	*Tim:*	You're doing seven again?
83	*Susie:*	Yep, I thought you asked me for seven.
84	*Tim:*	I said **it** doesn't work with seven. Is there another number like seven that **it** wouldn't work with?
85	*Susie:*	Oh, with forty, keeping the forty?
86	*Tim:*	I don't mind, you can change **it** if you want. I mean, is **it** the number you're dividing by that makes **it** work, or the number you're dividing into . . .
87	*Susie:*	I'm not sure if five does do **it** or doesn't do **it**. So could I find out if five does do **it**?
88	*Tim:*	Of course you can.
89	*Susie:*	But I'm not saying that this one will not work, OK? [Writes $30 \div 5$] I want to know whether **it's** doing **it** or not. [Pause] Six.

Table 2 Frequencies and ranks of words in the 'Susie' corpus and other selected language corpora

Word	Susie corpus		Pow corpus[a]		Howes corpus[b]		WCU7+[d]
	Per 1000	Rank	Per 1000	Rank	Per 1000	Rank[c]	Rank
you	34	1	25	5	19	6	53
and	29	2	29	4	38	3	1
the	29	3	38	1	41	1	2
it	29	4	23	6	17	8	6
a	28	5	32	3	24	5	3
that	25	6	16	8	15	10	68
I	24	7	32	2	40	2	4
to	20	8	21	7	25	4	5
of	17	9	11	10	18	7	19
one	14	10	9	12	5	18	23
so	13	11	4	23	9	11	62
is	12	12	5	21	7	13	7
what	10	13	5	20	4	21	143
with	10	14	4	22	5	19	14
be	9	15	6	18	5	17	90
do	9	16	9	13	4	20	104
in	9	17	12	9	17	9	11
number	8	18	0	25	0	25	630
would	8	19	2	24	2	23	168
yes	8	20	7	17	1	24	183
like	7	21	7	16	5	15	61
now	7	22	5	19	3	22	172
it's	6	23	8	15	7	12	492
this	6	24	8	14	5	16	40
on	6	25	10	11	7	14	12

Notes: a Fawcett and Perkins (1980): *b* Howes (1966). *c* Rank within this set of twenty-five words. *d* Edwards and Gibbon (1973).

> And then six lots of five; five lots of six [writes $6 \times 5 = 30$] . **It** does work.
>
> 90 *Tim:*   So **it** works with five.
> 91 *Susie:* With tens . . . ones, fives and tens **it** probably always works. And sometimes fifteens, twenties, twenty-fives, [then very fast] thirties, thirty-fives, forties, forty-fives, fifties, fifty-fives, sixties, sixty-fives, seventies, seventy-fives, eighties, eighty-fives, nineties.
> 92 *Tim:*   **It** works with all those numbers d'you think?
> 93 *Susie:* Sometimes **it** works. Sometimes.

The pronoun 'it' certainly appears to be prevalent in this passage. Indeed, the word 'it' appeared with greatest frequency (relative to transcript length) in the third of our five conversations. Table 2 indicates that 'it' is commonplace in all five of them. But in any case, how can one judge whether the occurrence[3] of the neuter third person singular pronoun in that corpus as a whole is in any way unexpected, or untypical? We could start with a list of words which children use frequently, such as that compiled by Rinsland (1945) in the United States, from children's writing and conversation. Studies of the vocabulary of English children include those of Burroughs (1957), from speech data, and of Edwards and Gibbon (1973), from children's spontaneous writing. Despite their differing methodologies, there is a high level of agreement between these two studies about the ranking of very common words. The speech/writing data in both studies were from children up to two years younger than Susie. However at age 7+ the eight words used most frequently are (in decreasing order of popularity): **and, the, a, I, to, it, is, my**. In fact the first five are way out in front on the basis of a 'popularity index'[4] used by Edwards and Gibbon, followed by 'it' and the other three, which are about equal to each other. Another, more recent, corpus of children's spoken language was collected by Fawcett and Perkins (1980) in South Wales. It consists of 65,000 words in 184 files, involving ninety-two children aged between 6 and 12, electronic versions of which have recently become available. Each child was recorded once in a play session and once in an interview. My analysis of the sub-corpus of 10-year-olds' speech indicates only slight variation in the 'top eight', compared with the data of Edwards and Gibbon: **the, I, a, and, you, it, to, that**. The emergence of 'you' in the Fawcett 10+ corpus will be seen to be of some significance. Table 2 shows the incidence of 'it' and of these other 'popular' words in my transcripts, aggregated over the full corpus. The table gives comparative data on these words with reference to other corpora, and on other words which occurred frequently in the transcripts. The table includes data from a broadly comparable study (Howes, 1966) of adult language.[5]

The pronoun 'it' (ranked sixth in the corpora of Edwards and Gibbon, and of Fawcett) is used more often in my conversation with Susie than these quantitative studies of natural language might lead one to expect, clearly ranking alongside 'and', 'the' and 'a'. Table 2 shows that only 'you' (which I will consider later) occurs more frequently, and even that is no longer the case if occurrences of 'it's are included.

Pimm (1987, p. 22) remarks that 'Like much informal talk, spontaneous discourse about mathematics is full of half-finished and vague utterances.' He proceeds to illustrate the point by drawing attention to the use of 'it' in an exchange (about enlargement) between a teacher and secondary pupil.

Teacher: [. . .] **Its** width is only a third as long as that one so – [. . .] how many of the smaller squares can you fill in? Nine, right.
Pupil: Is **it** that you square **it** – every time?

Pimm comments on the ambiguity of the referents for the occurrences of 'it', particularly in the pupil's question. He suggests that in saying 'you square it' the pupil is making a *generalization*. He picks up the 'crucial expression . . . *every time*' in evidence. Incidentally, the pupil's first 'it' could well be seen as an elliptic form of the cataphora 'Is it the case that . . .', drawing on the philosophical register.

Building on Pimm's observation, I have considered the varied purposes for which 'it' is being deployed in the corpus, and suggest that 'it' is a distinctive and important feature of maths talk, to the extent that it acts as a linguistic pointer, invariant at the surface level. The variable character of the referent of 'it' is illustrated in the examples which follow.

Our first conversation began as follows:

S1:1   *Tim:*   Imagine a square standing on its edge, on a table. Sketch what you see. [Susie draws the table and the square in 3D, showing edge of card]

2   *Susie:*   It's the width of it.

3   *Tim:*   Imagine a very thin square. The square rocks around on its bottom corners, jumps up, floats off the table, spins around, slows down, and drops back on to the table.

4   *Susie:*   It's all floppy over . . . it bounces and sort of flops down.

From the outset, in Susie's very first 'turns' (S1:2, 4) in our conversation, the 'it's are in evidence. There is a superficial and coincidental resemblance to Pimm's

Pupil: Is it that you square it – every time?

Squares feature in both situations, though one is algebraic (a product), the other geometric. The second 'it' in S1:2 and those in S1:4 are anaphoric as Susie makes reference to a square which I have already introduced in S1:1. Rather, she is referring to to her mental image of a square. With her first 'it' (S1:2) she refers to her drawing of the edge of the card; I am able to infer this from her gesture to the relevant part of the drawing. The pronoun is not co-referential with anything that has been (or is about to be) said.

This is an example of the linguistic phenomenon of deixis which was introduced in Chapter 4, by which a referent must be inferred by consideration of spatial, temporal, personal or other aspects of the *context* of speech. Thus (above) the pupil's first 'it' is cataphoric, formulating the question that follows, but the second is deictic (referring to a ratio of lengths).

Towards the end of the second session with Susie, I say:

S2:94   *Tim:*   One last thing. You remember last week we were doing multiplying by five, and you said that's easy? What you did, you multiplied by ten, you added a nought . . .

95   *Susie:*   . . . and then you halved it.

Here the referent is a number, ostensibly the original number multiplied by 10. Or is it the original number with a zero tacked on the end (which may be the same thing in form only; but recall my tale of Susie and the Problem of the Ten Rulers)? Or does economy suggest that it is precisely the original number which Susie halves, leaving the zero to hold the end place? Like the teacher's 'we', Susie's use of the pronoun 'you' – referring to herself but perhaps not only to herself – is an interesting instance of participant deixis (Wills, 1977); the 'you' has an indefinite, impersonal quality, to be considered later in this chapter.

The dialogue continues:

S2:98 *Tim:*   Now why, do you think, when you multiply by ten, you just add a nought?

99 *Susie:*   Because ... ten lots of ... If you count up in twos, suppose, the tenth will be twenty. And fours forty, fives fifty, and so on. Sixes sixty, sevens seventy, eights eighty, nines ninety, tens one hundred. And so on. Ten lots of, that's just a nought on the end.

100 *Tim:*   That's extraordinary, isn't it? Does that surprise you?

101 *Susie:*   Not to me.

102 *Tim:*   Why does that happen with ten, though, and not twelve, or nine, or ... Why does multiplying by ten add a nought on the end?

103 *Susie:*   Twenty would add two noughts, for instance.

104 *Tim:*   So if I multiplied, say, three by twenty, that would give three with two noughts?

105 *Susie:*   No. No, no, no, no, no. Silly me. Silly me. No. It's only with ten. For twenty you would double the number at the beginning.

106 *Tim:*   Why is ten the magic number like that?

107 *Susie:*   I don't ... it's just mathematics.

108 *Tim:*   [Laughs]. You mean the magic number might have been thirteen, or seven, but it just happens to be ten.

109 *Susie:*   By how the mathematics is made. How it was invented.

110 *Tim:*   Who invented it, then?

111 *Susie:*   No, I didn't invent it!

112 *Tim:*   No, *who* invented it?

113 *Susie:*   Don't know. They invented the mathematics and then people sort of added to it. It grew over the years.

114 *Tim:*   Why did the people that invented the mathematics make ten the magic number?

115 *Susie:*   I don't know. [Laughs]

In this case (S2:105, 'It's only with ten') the referent is a symbolic procedure. She is noting a property which belongs to powers of 10, but not to all multiples of 10. The property in question is clear from the context above. Indeed, I explicitly state the simplest version (S2:102); Susie formulates and subsequently withdraws a generalization of it.

I have included the concluding lines (S2:106–115) in which (109, 113) Susie makes anaphoric reference to 'mathematics', which she seems happy to discuss as an objective entity, even to the extent of giving it the definite article. Is it fanciful to suggest that Susie takes a non-platonist position *vis-à-vis* mathematics? She certainly sees it as something 'invented' rather than 'just there' or discovered. The clearest statement of her position is given quite spontaneously:

S2:113 *Susie:*   Don't know. They invented the mathematics and then people sort of added to it. It grew over the years.

I come now to some examples which show how Susie makes effective use of the pronoun 'it' to point to ideas of a general nature which neither she nor I have named, and whose nature I must be expected to infer.

Deixis

I shall demonstrate that Susie sometimes employs 'it' as a conceptual deictic, i.e. to point to concepts. The following example comes from our third session.

> S3:17 *Susie:* No, no . . . but times can do it can't it, and add, and take . . . no, take-aways can't do it.

The second 'it' seems to have the earlier 'times' (multiplication) as referent. The first (and third) 'it' is more problematic. On the basis of these eighteen words alone one can certainly surmise what the first/third referent – object of the verb 'to do' – may be. For example (given that Susie is nine), 'which operations make bigger'. Her syntax indicates that it is the operations themselves which can or can't do whatever-'it'-is. A more extensive context is required. A short lead-in gives some help:

> S3:12 *Tim:* Why is it that twelve divided by two is equal to six, then?
> 13 *Susie:* Well what it is, is this number [12] and see how many times that [2] goes into there [12]. How many times two goes into twelve.
> 14 *Tim:* Ahh . . . two goes into twelve . . .
> 15 *Susie:* Or twelve goes into two . . .
> 16 *Tim:* Or twelve goes into two.
> 17 *Susie:* No, no . . . but times can do it can't it, and add, and take . . . no, take-aways can't do it.

Susie has introduced the concept of commutativity into our dialogue. Not only does she not name the concept, but she is probably unable to give it a name; it would be surprising if she could. However, she certainly knows when 'it' holds. With the deictic 'it' she can articulate aspects of what she knows, and she does so quite spontaneously and unexpectedly.[6]

I regard her use of deixis in my final example (which is a prelude to the extended extract from transcript S3 given earlier in this chapter) as even more fascinating. To set the scene: Susie had divided 56 by 10 and written 5.6. She explained what 0.6 signifies:

> S3:28 *Susie:* It means it's six of the number you're doing it by, a tenth. Six-tenths.
> 29 *Tim:* Ah, six-tenths.
> 30 *Susie:* It's six-tenths of the real[7] number.

I'm suspicious that she is using decimals as remainders. To test this hypothesis, I ask her for 56 divided by 17. The question is a contingent one, chosen with the aim of precipitating cognitive conflict; it is chosen to retain the dividend (56) but to give 5 as the remainder. Susie duly writes 3.5. Closing in, I ask her:

S3:37   *Tim:*   What does that point five mean?
   38   *Susie:* It's a half of a . . . it's a half, of the real number. That's three of
                  the real number and the point five means it's going to be tenths.
                  The three is three whole numbers.

It now seems that her remainder-is-decimal rule is getting some interference from
her confident knowledge that 0.5 is a half. So next I set up 40 divided by 7. Susie
writes 5.5, and again confirms that the 0.5 is a half.

S3:40   *Tim:*   Right, let's get this absolutely clear. Is that five whole ones and a
                  half of one?
   41   *Susie:* Yep.
   42   *Tim:*   So how many lots of seven in forty? It's five and a half?
   43   *Susie:* Yep.

She continued, to my surprise, by volunteering an arithmetic deductive inference:

   43   *Susie:* Yep. So seven lots of five and a half is forty.

Susie has opened up a chink that I can exploit to 'correct' her general misconstruction
about decimals. So I press home the cognitive conflict, and ask her to work out
seven lots of five and a half (S3: 44–53). She obtains 38.5, preferring to work with
0.5 for a half.

S3:51   *Susie:* You could do this. [She writes $0.5 \div 7 = 3.5$, but corrects '$\div$' to '$\times$']
   52   *Tim:*   So what's seven lots of five point five altogether?
   53   *Susie:* Thirty-five point five . . . thirty-six point six . . . no, thirty-eight
                  point five. [Writes $5.5 \times 7 = 38.5$]

Thinking that I have a checkmate situation, so to speak, I home in:

S3:60   *Tim:*   Now what does that mean? [Indicates $40 \div 7 = 3.5$]
   61   *Susie:* How many times does seven go into forty.
   62   *Tim:*   When you actually work it out, five point five times seven, you
                  don't get the number you started with.
   63   *Susie:* I know
   64   *Tim:*   Isn't that a bit funny?
   65   *Susie:* No, that isn't, because whatever number you put in there [indic-
                  ates 7] you'd never reach forty, except for one, And you're not
                  allowed one.
   66   *Tim:*   Why not?
   67   *Susie:* 'Cos it's not really in the maths, it's just one, two, three, four,
                  five, six, seven, eight . . . [*Meaning the one-times table?*]

Susie is unperturbed. When 40 is divided by 7, and the quotient is then multiplied
by 7, she has no problem in living with a product which differs from 40, or so it
would appear. In fact, Susie subsequently enters into a lengthy and self-driven

exploration of what she considers to be 'special cases' in which divisor × quotient = dividend. First she considers 10, in which she has some confidence, but which she tests with a 'crucial experiment' (Balacheff, 1988, p. 218), taking 266 for the 'crucial' dividend.

S3:73	*Susie:*	If you do ten . . . and I think it would be ten if it . . . I don't, I'm not sure . . . suppose you had 266 . . . I'm not sure about this, I'll just find out.
74	*Tim:*	Yes, you experiment and find out. [Susie writes $266 \div 10 = 26.6$]
75	*Susie:*	That's twenty-six point six. And twenty-six point six lots of ten . . . so you'd in a way put a nought on the end, but you'd end up like that. [She has written $26.6 \times 10 = 266$]

Susie proceeds to consider whether 5 is also a 'special case', although her test case, involving 30, seems rather lenient in comparison:

S3:87	*Susie:*	I'm not sure if five does do it or doesn't do it. So could I find out if five does do it?
88	*Tim:*	Of course you can.
89	*Susie:*	But I'm not saying that this one will not work, OK? [Writes $30 \div 5$] I want to know whether it's doing it or not. [Pause] Six. And then six lots of five; five lots of six. [Writes $6 \times 5 = 30$] It does work.

She goes on to make conjectures, and she articulates generalizations freely:

S3:90	*Tim:*	So it works with five?
91	*Susie:*	With tens . . . ones, fives and tens – it probably always works. And sometimes fifteens, twenties, twenty-fives [then very fast] thirties, thirty-fives, forties, forty-fives, fifties, fifty-fives, sixties, sixty-fives, seventies, seventy-fives, eighties, eighty-fives, nineties.
92	*Tim:*	It works with all those numbers, d'you think?
93	*Susie:*	Sometimes it works. Sometimes.

One week later, at our next meeting, Susie is able to recall, with remarkable fidelity, what it was that she had come, in a tentative way, to believe:

S4:31	*Tim:*	OK, here's something left over from last week. We had $260 \div 10 = 26$ and $26 \times 10 = 260$. We also had $40 \div 7 = 5.5$ and $5.5 \times 7 = 38.5$ [Writes all these] OK? You remember that we talked about that?
32	*Susie:*	Yes, but if five or ten you do it with, it always comes out the same number.
33	*Tim:*	Yes, I was going to say that you said to me that sometimes. . . [interrupted]

34	*Susie:*	And sometimes fifteen, twenty, twenty-five, thirty, thirty-five, fifty, fifty-five, sixty, sixty-five, seventy, seventy-five, eighty, eighty-five, ninety.
35	*Tim:*	Ninety-five?
36	*Susie:*	Yes.
37	*Tim:*	A hundred?
38	*Susie:*	Sometimes.
39	*Tim:*	A hundred and five?
40	*Susie:*	Sometimes. Any by five or by ten will sometimes do it.
41	*Tim:*	What will it do?
42	*Susie:*	You start with the same number as you end.

Susie has abstracted a connection between dividing and multiplying which becomes the focus of so much of our subsequent conversation. She has no name for this relation, so she makes deictic use of the neuter third person pronoun, and frequently by saying that such-and-such a number (the divisor) will 'do it', or using the phrase 'it works' (S3:87, 89). Her vagueness achieves for her the goal of covering lexical gaps. Channell (1985, pp. 12–15) noted the same ability in her linguistics students in a tutorial, 'to get across a meaning where they do not have at their disposal the necessary words or expressions which they need to associate with the concepts they are forming' (p. 12).

Recall that Pimm notes the significance of the 'crucial expression . . . *every time*' in the formulation of his pupil's generalization. His (the pupil's) generalization is offered in the form of a question, giving it a certain tentative rather than assertive quality. Likewise Susie knows that there is a generalization waiting to be articulated, but she is uncertain about how comprehensive it can be. For convenience, denote by $s(b)$ the sentence 'For all n, $b \times (n \div b) = n$.' Now Susie is confident that $s(5)$ and $s(10)$ hold – this is evident in S4:32. Here, for Susie, the 'crucial' word is 'always', used with identical meaning and effect as 'every time'. In no way is S4:32 tentative. On the other hand she suspects the truth of $s(15)$, $s(20)$ and so on but conveys her doubt in the word 'sometimes', which she uses repeatedly, and in contrast to 'always'. One consequence of the maxim of Quantity is that 'sometimes' does not promise 'every time'.

This episode illustrates how *Susie's use of the deictic 'it'* enables us to share and discuss a concept which Susie possesses as a meaningful abstraction, yet is unable to name. This particular concept is an interesting case in point, since, I suggest, it has no name. I recognize, however, that words like 'inverse', 'reverse' and 'opposite' are commonly used in naming the sentence 'for all non-zero b, $s(b)$'. The Statutory Orders for the National Curriculum in England and Wales[8] opt for 'recognise that multiplication and division are inverse operations, and use this to check calculations'. I'm not happy with this statement – perhaps it is pedantry on my part? – because it is commonplace to conceive multiplication and division as binary operations (give me a pair of numbers, and I'll tell you their product). Thus, for me, the word 'inverse' implies, quite wrongly, that these ideas are set in the framework of a calculus of binary operations. But what we actually have (for every non-zero real number b) are two mappings (unary operations); multiplication *by b* and division *by b*; and now it *does* make sense to say that these are inverse mappings, under composition. It is well known (Graham, 1992) that the authors of the National Curriculum were in too much of a hurry to think about such niceties.

This analysis demonstrates that the beauty of the deictic 'it' lies in its function as conceptual variable. It (i.e. 'it') conveys the message 'I have something in mind. I know what I mean, and I think that you know what I mean.' It can be a linguistic pointer to a shared idea, to an understood but unnamed mathematical referent at the deep structure level. It can give both of us secure and economical access to an algebraic proposition, whilst Susie sets about trying to put bounds on its generality.

The notion of a (possibly tacit) object of attention is sometimes captured by the term 'focus'; see e.g. Chafe (1972) for a linguistic account, Garrod and Sanford (1982) for a psychological one. Moxey and Sanford (1993) connect pronoun use with 'focus', in a way that confirms my view of 'it' as a linguistic pointer to concepts which occupy the attention of the speaker:

> focus can be inferred through ease of pronominal reference. Because personal pronouns such as 'it', 'she' and 'they' carry only minimal information to recover the referent [. . .] it is clear that in practice things in a discourse that that can be referred to by pronouns must be a small subset of the possible previous antecedents, otherwise ambiguity would be rife. For this reason focus has become closely associated with the conditions of felicitous pronominal anaphora [. . .]. The point is that pronouns are good for referring to things in focus, while noun-phrases of a more complex and informative kind are best for things not in focus. In this way ease or acceptability of pronominal reference can be used *as an index and a probe for the state of focus*, other things being equal. (p. 58, my emphasis)

It was Susie who first caused me to notice and to reflect upon the deictic use of pronouns in maths talk. As a consequence of our first four conversations, over a period of a month or so, I began to work on her deployment of 'it'. There were signs, however, on my fourth visit to her school, that she was becoming irritated by my probing questions. I had asked her to explain her answer to the multiplication 8×32.4, which she had insisted on doing in her head. As she muttered to herself, I said:

S4:54　*Tim:*　　I hope you're going to explain this to me later.

I persisted until she said:

　　65　*Susie:*　　Help! . . . [pathetic tone] This will take *hours* to explain.
　　66　*Tim:*　　What will take hours to explain?
　　67　*Susie:*　　All of **it**.

Our conversation went into recess. When I returned, six months later, my attention shifted from 'it' to another pronoun.

On 'You' and Generalization

I have already noted (S2:95) Susie's use of 'you' to mean 'I' in saying '. . . and then you halved it'. Whilst such a use of 'you' as a vague referent is familiar in adults as well as children, it would make perfect sense (but imperfect truth) if Susie had intended 'you' to mean her audience (i.e. myself). The ability to interpret the deictic element (pronoun in this case) is dependent on knowledge of the 'co-ordinates' (time, place, speaker, topic, etc.) of the context. I now consider the particular significance of 'you' in mathematical conversation.

'You' can, of course, be used to address the person or persons to whom one is speaking. Such use is typically deictic but generally unambiguous. The following anaphoric examples come from longer transcripts in Brissenden (1988) of two groups of children using a computer program, Trains.

Craig:	Nine add seven, that's sixteen, it's one ten and six. [But he enters 6 first and then 1]
Steven:	Craig, you've got sixty-one now. It's the wrong way round . . .
Gavin:	Seventy-two plus seventy-two, that's a hundred and forty-four.
Teacher:	Gavin, that was quick. How did you work it out?

In adult–child mathematical conversations where the power relationship is asym-metrical, my transcripts indicate that the teacher/interviewer frequently uses 'you' to address the child, whereas the reverse is relatively rare.

It is as though such use of 'you' is a device which directly points to the other. And is it not rude to point? Use of the pronoun 'you' as a means of address can thus be interpreted as a message of power (Brown and Gilman, 1970; Hodge and Kress, 1988), inappropriate for the child in conversation with a teacher. My data strongly suggest that the majority of instances of 'you' by children in mathematics talk can be seen to be indicative of things that happen 'in the ordinary course of events' (Zandvoort, 1965, p. 128). Such things are *generalities*. Recall once more the pupil in Pimm's transcript:

Pupil: Is it that you square it – every time?

Recall Pimm's observation that one pointer to the fact that the pupil is offering a generalization is the expression 'every time'. Another, I suggest, is the use of the vague, unmarked 'you', functioning as a vague 'generalizer'.

In Chapter 4, I drew attention to this phenomenon in non-mathematical text. Here are four examples of it from my transcript data.

Anna

The extract which follows is a foretaste of data which will be introduced in the next chapter. Here, I am talking to two girls about ways of 'making' 20. Anna (aged 10) has proposed 'minus one add twenty-one'. In my 'bookend' questions, probing for Roksana's position, she (Roksana) is the referent of my 'you'. But now study Anna's 'explanation speech' (57), and consider 'Who is "you"?' for Anna.

T1:54	*Tim:*	Minus one add twenty-one. What do **you** think, Roksana? [Pause] Right, explain to us why that would give us twenty, Anna.
55	*Anna:*	'Cos nought add twenty equals twenty . . .
56	*Tim:*	. . . right . . .
57	*Anna:*	. . . so if **you**'re going into the minuses **you**'ve got to . . . em . . . **you**'ve, instead of saying twenty, that would equal nineteen, instead of twenty-one. And, minus, if **you**'re doing minus . . . one add, add minus one, something equals twenty, **you** go minus one add twenty equals, it equals nineteen. So **you** need to go minus one add twenty-one equals twenty.
58	*Tim:*	Are **you** convinced by that, Roksana?

In choosing the impersonal 'you' in preference to 'I', Anna has 'decentred' and become, in some sense, detached from what she is asserting. Personal confidence – albeit tentative – in the general application of some process or proposition enables the speaker to offer it for others to appropriate. When Anna says (T1:57), 'you go minus one add twenty equals, it equals nineteen', it could be that she is sharing some kind of number line imagery that she believes should be accessible and convincing to others. In any case she is saying that 'anyone can do this'.

Simon

I began the first of three mathematical conversations with Simon with an enquiry along the lines 'Give me two numbers whose sum is ten'. I have begun undergraduate courses in Number Theory with the same question – intended, in that situation, to explore what 'number' means to the students. After similar preliminaries with Simon, he proceeded to determine the number of ways that any positive integer could be 'made' as a sum of two positive integers. Some time later, I adopted 'Make Ten' for the next phase (Chapter 6) of contingent interviewing with a larger sample of children.

Simon progressed to consider making a given integer as a sum of three integers. First he worked on 20 and came to see that the number of ways is the eighteenth triangular number. Picking up the conversation at that point:

Si1:205	*Tim:*	Right. Suppose instead of twenty, right, I said how many different ways are there of adding up three numbers to make fifty?
206	*Simon:*	I'd do forty-nine times [pause] twenty-five.
207	*Tim:*	[pause] You'd better explain that.
208	*Simon:*	Um, no, I wouldn't. I'd do forty-nine times twenty-four.
209	*Tim:*	Explain it.

210	*Simon:*	Well, it's the triangular number of, em, it's the . . . working out the triangular number of . . . the forty-eighth triangular number.
211	*Tim:*	Mm-hm. Why do you know that?
212	*Simon:*	I'm going on the assumption that it works the same for twenty.
213	*Tim:*	What happens with twenty?
214	*Simon:*	It, em, I found the triangular number for eighteen, because . . . the second number before twenty.
215	*Tim:*	Right, right. So what you do with fifty, you say . . .
216	*Simon:*	Make, work out the triangular number forty-eight.
217	*Tim:*	Right.
218	*Simon:*	And to do that, I times it by . . . so I do forty-eight times – no, I do forty-nine times half of forty-eight, which is twenty-four.
219	*Tim:*	Right. Can you see why it's forty-nine times half of forty-eight?
220	*Simon:*	Yeh.
221	*Tim:*	Why?
222	*Simon:*	Because, to work out a triangular number, you get the first and the last, and the second and that . . .
223	*Tim:*	And multiply it by how much?
224	*Simon:*	Um, the num . . . a half of the number . . . of . . . half the number of numbers you've got. So it's like from nought to forty-eight, so half of that, 'cos you've only got half the numbers to work out.

Observe in passing Simon's expression 'it works' in Si1:212, strongly reminiscent of Susie's use of 'it' to point to a general relationship or procedure. In 206–212 he uses 'I' to describe what he did for 20 and what he predicts for 50. Whereas I use the vague generalizer 'you' in 215, Simon persists with the personal 'I' in 218 to describe how he would calculate a particular triangular number. In 222 and 224, however, he adopts the pronoun 'you' himself: here he is formulating a *general* procedure for such a calculation, and explaining why it works – Gauss's method, in fact, which I had in effect introduced him to earlier, for the purpose of finding the eighteenth triangular number.

Susie

My fifth and final conversation with Susie was memorable. Elsewhere, I have given the remarkable *mathematical* content the attention it deserves (Rowland, 1997a). In this extract, Susie is developing a highly idiosyncratic method of dividing 100 by various fractions, five-sevenths in this instance.

| S5:62 | *Tim:* | OK. Now, you said that it wouldn't work for seven-ninths, didn't you, this method? Right? Now, I'd just like you to write down five-sevenths, just here. |
| 63 | *Susie:* | I'm going to have to think, though, very well. Um, I'll try . . . [pause]. Ahh, of course . . . [interrupted] |

> 64   *Tim:*   You have a think while I push the door up.
> 65   *Susie:*   ... You can't ... I don't understand. It's definitely a hundred. So that means two ... Ahh, ahhh [*big moment*] you've got two left, and you need five each time. So if you have two hundred ... um ... divided by five. How many times does five go into two hundred? Well, it goes into one hundred twenty times ...
> 66   *Tim:*   Mm-hm.
> 67   *Susie:*   Must go into forty times. So that's ... a hundred and forty.

Notice how Susie's 'I' (63, 65) becomes 'you' (65) after the 'ahh' which seems to signify the moment of insight.

Again, the pronoun 'you' is an effective and non-trivial pointer to a quality of thinking. Susie's shift from 'I' to 'you' signifies her reference to a mathematical generalization.

The generality expression 'each time' occurs in (65) as part of an account that has the quality of a generic example, for the exposition of the division method that Susie is developing.

In my corpus of 25,000 transcribed words of mathematical conversations with 9 to 11-year-old children, 'you' is the most frequently used word (744 occurrences), followed by 'and' (662), 'to' (400) and 'a' (394).[9] In the 'dividing by fractions' conversation with Susie above (2,450 words), Susie and I each use 'you' forty times. *Every* time I use the word I am addressing Susie, whereas she uses 'you' to address me *only twice* – once to make sure that she's not leaving me behind!

> S5:187   *Susie:*   And it has to be two hundred. So you would have two hundred [writes] divided by five. Do you understand that?
> 188   *Tim:*   Yes, thank you.

On the whole, Susie reserves 'I' to mark her feelings and beliefs, or accounts of her personal actions, whereas 'you' indicates a kind of detachment from her strategy and computational methods.

The same 'personal versus general' markers are evident in the final example.

Katy

In an undergraduate supervision with me (NT7), Katy is talking about her progress with a project on continued fractions:

> NT7:17   *Tim:*   And what have you proved?
> 18   *Katy:*   Um, I've proved that ... I think, now I want you to have a look at it, 'cos I'm not sure if it's right, but I did this [...] yeah, I was trying to, I've had a look at, I said that, right, the root of a squared plus one is equal to a plus one over alpha, like we did before ...
> 19   *Tim:*   Oh yeah, yeah, yeah, go on, yes.

20 *Katy:* Um, and it's also equal to *a* plus one over two *a* plus one over
         alpha because **you** can, because alpha goes on for ever **you** can
         start it whenever **you** want.
21 *Tim:*   Umm . . . yeah, go on.
22 *Katy:* And so **I** put those two equal to each other.

The occurrences of 'I' are all set within accounts of personal actions located in time and space. This is also true of the 'we' in Katy's first turn (18). I suggest, however, that the 'you's in her second turn (20) are located in some informal object language which is appropriate to the specification of procedures and algorithms – because 'you' is not any particular, actual person, rather it is anyone. Again, the pronoun functions as a vague 'generalizer'.

Summary

My interest in pronouns in mathematics talk, and my observations about them in this chapter, derive from my belief in generalization is the mathematical process *par excellence*. This conviction coincides with and relates to Mühlhäusler and Harré's view (1990, p. 58) of the primacy of the deictic role of pronouns. Analysis of the use of pronouns in the mathematical discourse of novice users of the mathematical register reveals:

- deictic use of 'it' to refer and point to mathematical concepts and generalizations which have not (or, for various reasons, cannot) be named in the discourse;
- the pronoun 'you' as an effective pointer to a quality of thinking involving generality; the shift from 'I' to 'you' commonly signifies reference to a mathematical generalization.

The use of 'it' as a conceptual deictic enables the pupil to say what s/he could not say otherwise, to draw attention to mathematical entities whose name s/he does not know. The notion of 'focus' as locus of attention (Moxey and Sanford, 1993, p. 58) is important here, for the teacher who is sensitive to the pronoun/focus connection can be made aware of the presence of a cognitive focus involving generalization.

The second person pronoun 'you' is a prevalent and effective pointer to generalities in mathematical discourse. This is perhaps especially true in children's discussion with their teachers, a context in which children rarely address the adult participant as 'you'. Like Delderfield's bank clerk (Chapter 4), pupils in mathematics classrooms need ways of distinguishing, for their audience, their experiences and feelings from detached observation and generalized objective comment. These two qualities coexist in mathematical activity, and both are necessary.

The theme of generalization continues in the next chapter, but the emphasis shifts to pupils' pragmatic means of conveying propositional attitude as they assert predictions and generalizations.

Notes

1   A list of conventions and transcripts can be found in the appendices.
2   On the other hand, Wales (1980, p. 33) accounts for the prevalence of 'we' in scientific discourse in terms of the egocentric force of pronouns in English. Thus the choice of 'we' in preference to 'I' is made in order to achieve rhetorical distancing of the speaker/writer from the content of what s/he says/writes, to achieve muted egocentricity.
3   The value of counting particular words in transcripts is limited, but electronic text storage now makes such counts both possible and relatively quick.
4   The 'popularity index' used by Edwards and Gibbon is defined as the product of the percentage of children in the sample using the word and the average use per child in the sample, e.g. if precisely half the children in the sample used a particular word, and each of them used it just once, the FI would be 25 (50×0.5). At age 7+ the highest popularity indices were found to be 3,667 ('and'), 2,968 ('the'), 2,349 ('a'), 2,330 ('I'), 1,915 ('to'), 943 ('it'), 913 ('is'), 889 ('my'), 842 ('go').
5   For comparison, I have consulted adult language data from the Lancaster–Oslo–Bergen (LOB) project (Hofland and Johansson, 1982) and Howes (1966). The LOB corpus is about 1 million words of written language from fifteen categories of British sources (fiction, newspapers, etc.), all published in 1961. Howes's corpus consists of 250,000 words spoken by American adults. The relative frequency and ranking in the LOB corpus of the twenty-five relevant words in the LOB corpus are shown in Table 2. Howes does not list rankings, but relative frequencies are shown in the table together with rankings of the thirty words among themselves. As I have remarked, the LOB and Howes data are from adult sources and must be used with caution in the present context. In fact (Table 2) they support the general conclusion that one would not expect 'it' to occur nearly as frequently as 'the', 'and' and 'to' (ranked 1, 2 and 4 in both sources). Howes also indicates that 'I' is generally much more prevalent than 'it' in (adult) spoken language, although, not surprisingly, the reverse is the case in adult writing (LOB).
6   Susie's use of the deictic 'it' to refer to things she cannot name has a parallel in concern for propriety in adult social practice, giving rise to deictic reference to semi-taboo topics, cf. the car sticker 'Windsurfers do it standing up' and numerous variants. The *double entendre* is achieved by the exploitation of vagueness.
7   Susie's use of 'real' to mean 'whole' is quite a discussion starter. It is certainly reminiscent of Kronecker's famous remark that 'God made integers, all else is the work of man', and calls to mind her comments about 'the mathematics' being invented.
8   Quoted from the second iteration (DES, 1991) of the Mathematics National Curriculum for England and Wales. The third, post-Dearing, version (DFE, 1995) has in its Level 5 description for Number and Algebra (p. 33): 'They check their solutions by applying inverse operations or estimating . . .'
9   For comparison (Table 2) in the Fawcett corpus (Fawcett and Perkins, 1980) of 10-year-olds' speech, 'you' is ranked fifth (averaging twenty-five occurrences per 1,000 words, compared with thirty-four in the 'Susie' corpus) after 'the' (thirty-eight per 1,000), 'I' and 'a' (both thirty-two), and 'and' (twenty-nine). 'It' comes next (twenty-three), as it happens.

6   Hedges

Policeman: How many times did you hit him?
Ash: A couple of times, maybe.
(*Casualty*, BBC1 TV, 5 November 1994)

Mathematics, viewed as a field of human endeavour, as opposed to an inert body of knowledge, offers considerable potential for intellectual risk-taking. To participate in this endeavour we are obliged to lower defences, and in speaking, to expose ourselves. Nothing ventured, nothing gained; encouraging pupils to take risks, in the form of making predictions and conjectures, might be expected to be a feature of effective teaching. Of course, there are also risks associated with public display of knowledge of any kind, even at 'low levels' of mathematical activity – recall of facts, rehearsal of algorithms – especially for those who have suffered ridicule at the hands of insensitive teachers or inconsiderate peers. Mathematics holds an invidious reputation within the school curriculum, being associated with fear of error and consequent public humiliation. The scars which persist into adulthood were brought sharply into relief by Sewell's study (1985) of adults' use of mathematics in daily life.

One common perception is that the questions teachers ask their pupils are not searchlights focused to reveal truth, but traps set to expose ignorance. Janet Ainley, studying children's perceptions of the purposes of teachers' questions, calls such uses 'testing questions':

> Because testing questions are so common, particularly in mathematics where answers are seen as being clearly 'right' or 'wrong', there is a danger that pupils may perceive all teacher questions in this way. Such a perception would inevitably be detrimental to attempts to encourage discussion, investigative work or problem solving in mathematics: pupils will feel that the teacher always knows the 'right' answers to any questions she asks, and furthermore that the teacher is always judging pupils by the answers they give. It is not surprising that pupils are reluctant to risk giving 'wrong' answers in these circumstances. (1988, pp. 93–94)

From a speech act perspective, Labov (1970) comments that a question is normally deemed appropriate only when the enquirer meets certain speech act 'sincerity' conditions – including, in this case, that s/he doesn't know the answer, would like to know it, and has reason to believe the hearer is able to supply it. Each of these conditions would be satisfied when, for example, you make a telephone enquiry to the railway enquiries office. Labov shows that questions in classroom situations are exempt from these rules, and that the conditions governing appropriateness in the answers to such questions differ accordingly. The expectation of pupils is that 'classroom questions' (R. Lakoff, 1973) are not genuine requests for information, but public 'requests for display' (Labov and Fanshel, 1977). These differ from other questions in that the enquirer (A) already has the information sought in the question, and 'the request is for B to display whether or not s/he has the information' (*ibid.*, p. 79). Given the privileged position of the enquirer in the dialogue, the 'display' of the respondent is a revelation of knowledge (public success) or ignorance (public failure).

Laurie Buxton has addressed the issue of success and failure in the mathematics classroom from the emotional perspective of the learner:

> Most classroom maths sets tasks, often with very clearly defined goals; whether they have been reached or not is seldom in doubt [...] This clarity tends to enhance the sharpness of emotional response. There is a nakedness about the success or failure in reaching a goal that evokes clearly defined emotions whose nature one cannot disguise to oneself. (1981, p. 59)

Anne Watson recognizes how, in time, this can generate inhibitions in pupils, who:

> are worried about being wrong and nervous about asking for help if 'being wrong' and 'needing help' have, in the past, been causes of low self-esteem by leading to ridicule, labelling or punishment. (1994, p. 6)

This tendency is illustrated in the words of two 16-year-old girls (C and S) who had chosen not to continue mathematics studies at school. Here they are interviewed by Susan Hogan, who asks about their experience of 'speaking out' in mathematics classrooms:

> *SH:* And do you need to be confident in order to speak . . . ?
> *C:*   Yeah, because there were people in the class that were so good that you kind of –
> *S:*   – thought well, they're gonna laugh at me if I get it wrong.

(From a presentation at the Open University, 11 November 1995)

Whilst Buxton speaks of nakedness, the word that comes to me is vulnerability. Not, of course, a state of being in danger of physical offence, but exposure to intellectual injury. How odd, how unfair it is that the 'crime' is cognitive but the penalty affective. Extrinsic and intrinsic sanctions are associated with being wrong; it follows that there is a high premium attached to being right, with insufficient acknowledgement by pupils and teachers that uncertainty is a common and valid, indeed an honest and honourable, state to be in. One could go further, and insist that uncertainty is a productive state, and a necessary precondition for learning. For once we believe we 'know', we are no longer open to the possibility of further knowing. When mathematics is coming into being in the awareness of an individual, uncertainty is to be anticipated and expected. This is the essence of Mason's 'conjecturing atmosphere':

'let it be the group task to encourage those who are *unsure* to be the ones to speak first [. . .] every utterance is treated as a *modifiable conjecture*!' (1988, p. 9)

The absence of such an atmosphere in the experience of the students interviewed by Hogan (above) is evident:

SH: Would you ever volunteer an answer in class [. . .]?
S: Yes, if it's something I definitely know the answer to I'll put my hand up and say the answer.
SH: And if you're not sure?
S: I don't – I wait for someone else.

Yet, in the making and learning of mathematics, uncertainty ought to be expected, acknowledged and explicit. All learning – in the sense of coming to know and believe something that was formerly unknown or doubted – involves some kind of act of commitment on the part of the learner.

In Chapter 2 I have described generalization as a manifestation of inductive reasoning, as a kind of cognitive leap, embracing enthymematic premises that cannot be *known* to be true at the moment when a conjecture is made. To assert such a conjecture is indeed a risky business. I have never found a way – for there can be none – of teaching investigationally without risk-taking consequences for my students, and sometimes for myself. Learning investigationally, by action and reflecting on action, is by no means universally welcomed by pupils, many of whom feel that the risk involved is too high a price for the *frisson* of discovery. The articulation of the predicting or generalizing insight – making public what it is that one has 'seen' – can be shot through with uncertainty until it meets with approval: David Wheeler has written:

Investigations have to be 'managed' sometimes, as in 'I managed to solve the problem'. [. . .] It is worth asking about students who have not managed their anxiety. Managing an investigation certainly involves not only managing the technical and mathematical tools, but the affective components too. Solving problems and investigating situations (and even mastering conventional mathematics) are risk-taking activities and require courage as well as skill. It would be good if some writers (without being sentimental about it) would give some attention to these often unspoken aspects. (1984, p. 25)

Given a commitment to that style of teaching, through posing problems and promoting investigations, one must face up to the management of affect. The task becomes one of creating a conjecturing atmosphere, stimulating but trusting and accepting. For many years, mathematics teachers in the UK have been fortunate to be sanctioned to ride a tiger in our classrooms! 'Investigations' – a wild animal so enthusiastically nurtured by the ATM in the 1960s, recognized and approved by Cockcroft in 1982, and finally adopted, and domesticated by the GCSE examining boards.

In a quasi-empirical environment for mathematics teaching and learning, the process of 'coming to know' is an exhilarating ride – at times risky, infused with uncertainty, at other times replete with 'approaching completeness' (Watson, 1995). In this chapter I describe and analyse some ways in which uncertainty is coded in spoken language, with reference to a mathematical study carried out with children aged between 10 and 12. I chose to work with them on a task which required (among other things) prediction, generalization and explanation. Linguistic pointers to such moments of uncertainty will be identified. I shall draw attention to the presence of hedges in pupils' mathematical discourse, and analyse how speakers do things with them.

Method

The study described here arose from my interest in the language that children (specifically those in the 9 to 12 age range) use to invoke, describe and engage with the mathematical process of generalization. Following the case studies with Susie and Simon, I moved to replication of an 'experiment' with a number of children. Intending to diminish my contribution to the mathematical discussions, I worked with ten pairs of children aged 10 or 11 for about thirty minutes with each pair, and encouraged some peer interaction on a common mathematical task. I did, however, remain as a participant, as opposed to a passive observer, originally so as to maximize their engagement with the mathematical task.[1] Contrary to my original intentions, I ended up reflecting on the pupil–interviewer interaction too. This experience parallels that of Paul Cobb and his collaborators (1992) as reported in their account of a teaching experiment with a class of second-grade children:

> we initially took a radical constructivist position [. . .] and focused almost exclusively on students' learning from a cognitive perspective. Only later did we widen our purview to encompass an interest in the teacher's learning and classroom social interactions, and begin to complement our initial constructivist position with constructs derived from symbolic interactionism [. . .] and ethnomethodology. (p. 101)

The interview technique which I deployed is a mildly restricted form of the contingent questioning (Ginsburg, 1981; Ginsburg et al., 1983) which I described in Chapter 1. Recall two of the features of the contingent interview, a method which:

- employs a task to channel the subject's activity;
- has some degree of standardization.

Choice of task. I compiled a number of tasks and considered their competing merits in relation to certain requirements – the chosen task would be replicated with ten pairs of children, I could not ditch it mid-stream. It had to be accessible mathematically to most 10-year-old children; have some intrinsic interest for them; have definite potential for stimulating prediction and generalization; preferably be accessible in terms of their explaining their generalizations; and must reasonably be expected to lead to some worthwhile (if unpredictable) mathematical outcomes in about half an hour. In the end, I found that 'Make Ten' – an easily accessed combinatorial problem – comfortably met these requirements. The idea of the task originated in my teaching and was piloted in one of my earlier weakly framed conversations with Simon.

Standardization. The mathematical encounter with each pair of children was standardized to the extent that I intended each 'Make Ten' interview to proceed in five 'phases' which I had planned in advance. I now describe the task and indicate its component phases.

I initiated each interview with the same conversational gambit[2] – inviting consideration of the number of ways that 10 can be 'made' as a sum of two 'numbers'. This combinatorial problem will be familiar to primary (elementary) school teachers, although its potential as a starter for making generalizations may not. I did not work from a 'script', but the following opening (with two boys, Jubair and Shofiqur) is typical.

T5:1	*Tim:*	Jubair, I'd like you to give me any two numbers that add up to ten.
2	*Jubair:*	Six add four.
3	*Tim:*	Six add four. Shofiqur?
4	*Shofiqur:*	Eight add two.
5	*Jubair:*	Five add five.
6	*Tim:*	OK, so you get the idea. Now what I want you to decide between you is how many ways is it possible to do that?

The next three exchanges were not at all typical, however:

7	*Shofiqur:*	[almost instantly] Nine ways.
8	*Tim:*	Nine ways.
9	*Jubair:*	No, ten ways.

It was more usual for the children to list the possible sums, orally or on paper, and then to count them. Then I would say something like:

Now just as you eventually decided about that question for ten, I'd like you to decide between you how many different ways are there of doing that for twenty?

The first phase of the interview was planned to proceed as above with similar examples of listing sums and counting how many had been found. I would propose the numbers to be 'made' in this way, my choices depending on the children's earlier responses to my questions about 'making' 10 and 20 – in particular, on the facility they displayed and whether reversals such as $2 + 8$ and $8 + 2$ were both counted.[3] The next phase would then involve my proposing a further target number – say 30, 50 or even 100 – slightly out of the range of those already counted, and inviting a prediction of the number of ways this number could be made.

T3:60	*Tim:*	OK. Now a funny question. Supposing the number now that I'm interested in is twenty-three? So you've got twenty-one add two, and so on. How many ways?
61	*Caroline:*	Twelve, would it?
62	*Alex:*	Yeah, probably.
63	*Caroline:*	Twelve.
64	*Tim:*	What makes you think twelve, Caroline?

Subsequent phases, contingent on preceding ones, would involve my probing for the thinking behind this prediction (and possibly others) and discussion of per-ceived 'rules' – conjectures about what might happen with 'any' number. For example:

| T4:138 | *Tim:* | Right, OK. Is there a kind of rule that you could state gener-ally? I mean supposing I now picked out any number . . . you know, like five hundred and thirty-seven or something . . . and said how many ways can you make that from adding two numbers. How would you know what the answer was? |
| 139 | *Alan:* | Just take away one, and then you'll know how many you can get. It's the same here. Ten, there was nine possibilities; twenty, there was nineteen possibilities; thirty-seven, there was thirty-six possibilities. |

In some cases we continued to test the generality of such conjectures, and tried to see why they might be true 'in general'. In practice, such proofs were always founded on the possibility of 'seeing the general in the particular' (Mason and Pimm, 1984), producing confident awareness of how things would be for any other particular, as it were. That is to say, the pupils explained their generalizations by accounts of generic examples. One instance (a lengthier quotation from Alan and his partner Harry) is given at the end of this chapter.

Restriction. In an extreme, weakly framed form of contingent interviewing, no contribution is deemed irrelevant; the interviewer follows the thoughts of the inter-viewee and encourages him or her to develop them further. In this sense the subject determines the agenda and is free to deviate from it or to redefine it. My extended conversations with Susie and Simon were of this kind. The 'Make Ten' interviews were intended to generate some data as a basis for prediction and generalization. The interviews were contingent in the sense that the children were given freedom to decide what kind of sums would be eligible, and which sums would be counted as 'different'. On the other hand, I expected to shape each conversation as indicated above, aiming for prediction and generalization. Each of the 'Make Ten' interviews was audiotaped and transcribed – a corpus of some 25,000 words.

'Make Ten': Frances and Ishka

The following extracts give a fuller picture of how conversations arising from the task developed. They are from an interview with two girls, Frances and Ishka, both about 10½ years old. Here, I have separated the transcript (T6) into five 'episodes', A to E. These 'episodes' are thus the realized equivalents of the planned 'phases' of the interview. For most of the interviews the phase–episode correspondence was fairly close. The episodes of T6 are summarized and illustrated by the following extracts.

Episode A [7½ minutes]. In this phase I introduce the problem, that of making 10.

The children list and decide five ways, allowing no reversals (decided by Frances).

T6:16	*Frances:*	Four and six, five and five, six and . . . oh, that's the same.
17	*Ishka:*	Five ways?
18	*Frances:*	Maybe.
19	*Ishka:*	Mm, maybe . . . I think . . .
20	*Frances:*	What do you think?
21	*Ishka:*	We haven't had five, five have we?
22	*Frances:*	We have!
23	*Ishka:*	Oh, OK, er . . .
24	*Frances:*	The others are like if you do six four, we've already done four six.
25	*Ishka:*	Mm [sighs].
26	*Frances:*	Shall we just say five ways?
27	*Ishka:*	There's about five.

I then ask about making 20. They list and decide ten ways. Finally, I ask about making 13; Ishka lists and decides six ways.

Episode B [2 minutes]. I begin by asking them to recall results so far. Then I ask about making 30, inviting an initial prediction. Frances predicts 15, Ishka agrees, and explains how her prediction relates to the earlier results. There is an air of plausibility rather than certainty in their attitude to Frances's prediction.

T6:105	*Ishka:*	I think there'll be around . . .
106	*Frances:*	Fifteen?
107	*Ishka:*	Yup.
108	*Frances:*	Maybe? [. . .]
113	*Ishka:*	Most of them are half or just about one away from . . . [. . .]
117	*Tim:*	OK. Prediction of fifteen. Er, so . . . let's just go back to what Ishka was saying. She was saying that in most cases it's about half.
118	*Ishka:*	Well, yes, 'cause ten was five.
119	*Tim:*	Right. [. . .]
120	*Frances:*	Twenty was ten. [. . .]
123	*Frances:*	Thirteen was about six.
124	*Ishka:*	But, er, thirteen was six.
125	*Tim:*	OK.
126	*Ishka:*	Although that isn't exactly half.

I proceed to ask about making 100. Frances instantly answers, 'Fifty?' Ishka agrees in a vague way, but conveys considerable uncertainty when pressed to commit herself to Frances's prediction.

129	*Frances:*	Fifty?
130	*Ishka:*	About fifty, yeah.
131	*Tim:*	About fifty. Now are you saying about fifty, Ishka, because you're sort of playing safe or, I mean, do you really think it is fifty?
132	*Ishka:*	Well maybe not exactly, but it's around fifty basically?
133	*Tim:*	OK. And Frances, do you think it's exactly fifty or around fifty?
134	*Frances:*	Maybe around fifty.

Episode C [3½ minutes]. I return to making 30 – how sure are they that there are 15 ways?

135	*Tim:*	OK. What about this prediction of fifteen for er thirty, was that around fifteen or exactly fifteen?
136	*Ishka:*	It's fifteen or around.
137	*Frances:*	Yes.
138	*Ishka:*	'cause we can't be exactly sure until we've tried it, but . . .

The girls list and count fifteen ways.

Episode D [1½ minutes]. Looking for some structure in the list for 30:

194	*Tim:*	[. . .] you started at the top and went one, two, three, four, five . . . Can you notice something about these numbers that would immediately tell you there's fifteen without writing them all down?
195	*Frances:*	There's one to fifteen, so fifteen numbers?

Then imagining what a list for 100 would look like:

198	*Tim:*	Right. If you did a similar listing for, um [. . .] a hundred [. . .] If you did a list like this and you started at . . . what would be the two equal numbers at the top?
199	*Ishka:*	Fifty and fifty.
200	*Tim:*	Fifty add fifty, and then it would go forty-nine [. . .]
209	*Tim:*	And how many numbers would there be in those, or how many ways would you have done it?
210	*Frances and Ishka:*	Fifty.

Episode E [4 minutes]. Articulating general 'rules' for the number of ways:

212	*Ishka:*	So the even numbers are exactly half.
213	*Tim:*	That's really good, Ishka.
214	*Ishka:*	And the odd numbers are . . .
215	*Frances:*	Are exactly half.
216	*Ishka:*	Are sort of, are one off . . . that.

Generic examples suggest proofs of the rules:

244	*Tim:*	Right, OK, I just wonder whether it's possible by writing down all the ways of getting, say, thirteen like this . . . yuh? . . . whether you could then see why it was half of the number before. [. . .]
245	*Frances:*	Yes.
246	*Tim:*	Do you want to try it, Frances?
247	*Ishka:*	So we start at half of thirteen.
248	*Frances:*	Half of thirteen would be . . .
249	*Ishka:*	Six.
250	*Frances:*	Six and seven. Six point five . . . [laughs]
251	*Tim:*	You can't have six and a half and six and a half, so what will you start with?
252	*Frances:*	Six and seven.

We discuss why there must be six sums in the list: the interview concludes.

Hedge Types in Mathematics Talk

Each of the four categories of hedge described in Chapter 3 is in evidence in the Make Ten transcripts, and is associated with particular kinds of goals. Here I take a first look at the pragmatics of these categories in mathematics talk, with some examples from Make Ten.

The four basic types

Plausibility Shields are typified by 'I think', 'maybe' and 'probably', as in this excerpt from the episodic overview of the Frances/Ishka 'Make Ten' interview.

T6:16	*Frances:*	Four and six, five and five, six and . . . oh, that's the same.
17	*Ishka:*	Five ways?
18	*Frances:*	**Maybe.**
19	*Ishka:*	Mm, **maybe . . . I think . . .**

A Plausibility Shield implicates (in the Gricean sense) a position held, a belief to be considered – as well as indicating some doubt that it will be fulfilled by events, or stand up to evidential scrutiny. It is a means of offering an idea without the obligation of commitment to its truth.

The second kind, an *Attribution Shield*, implicates some degree, or quality, of knowledge to a third party. In the Make Ten data there are relatively few Attribution Shields, and they tend to be used by me rather than the children, as a teacher-like device for meta-comment (Pimm, 1992) on the activity. Thus, with Kerry and Runa:

T8:176	*Tim:*	OK. Um, how many ways would there be, then, for twenty-four?
177	*Runa:*	Twenty-four? Add?
178	*Tim:*	Same kind of thing, but with twenty-four.
		[Kerry is whispering, seems to be counting something]
179	*Runa:*	Um, nineteen. Nineteen ways.
180	*Tim:*	Nineteen ways, **says Runa**.
181	*Runa:*	I just guessed.
182	*Tim:*	Kerry's still thinking.
183	*Kerry:*	Ten.

Here I use Attribution ('says Runa') in T8:180 in order to feign neutrality on my part about the contribution of one child, in order to obscure my evaluation of her answer (T8:179), and to encourage the participation of the other child.

The semantic effect of *Approximators* – the second major category of hedges – is to modify (as opposed to comment on) the actual proposition, making it more vague. 'About', 'around', and 'approximately' are examples of Rounders, which constitute the first subcategory of Approximator. Rounders are usually associated with estimation in the domain of measurements, of quantitative data. Association with prediction and generalization does not readily come to mind, yet Rounders occur frequently in the Make Ten corpus, to qualify combinatorial prediction, as in Episode A with Frances and Ishka:

| T6:26 | *Frances:* | Shall we just say five ways? |
| 27 | *Ishka:* | There's **about** five. |

and again, in Episode B:

T6:105	*Ishka:*	I think there'll be **around** . . .
106	*Frances:*	Fifteen?
107	*Ishka:*	Yup.

Adaptors, such as 'a little bit', 'somewhat', 'fairly', attach vagueness to nouns, verbs or adjectives. The following examples are from the Make Ten interview with Jubair and Shofiqur. Shofiqur has just indicated what a list of ways of 'making' 20 would look like, and predicts twenty-one different ways.

| T5:66 | *Shofiqur:* | . . . It's just **a bit** the same, like this [indicating the list for 10]. |
| 67 | *Tim:* | So Shofiqur is **pretty** convinced that it's twenty-one. Right, are you persuaded by his argument? |

68	*Jubair:*	Not **really**.
69	*Tim:*	Have a go at – I'm **fairly** convinced what you said, Shofiqur, have a go at convincing Jubair that there are twenty-one ways. I mean, take it slowly.
70	*Jubair:*	Come on, then!
71	*Shofiqur:*	I only took a guess.

Adaptors suggest, but do not define, the extension of categories, concepts and so on. Thus Shofiqur uses an Adaptor phrase 'just a bit' with respect to same(ness); I myself use two Adaptors here, 'pretty' and 'fairly' (67, 69), to suggest, first that Shofiqur's conviction, then mine, is not simple and unreserved, but of a fuzzy kind.

A sift of the transcripts suggests that it is I, rather than the children, who make most use of Adaptors. Like Attribution Shields, and for similar reasons, I use them as a means of commenting on the children's contributions. Specifically, I use them to make indirect comments on their predictions, generalizations and explanations.

The taxonomy provides a setting for studying the significance of the various hedges used in my Make Ten interviews. The framework is useful in drawing distinctions and providing starting points. Whilst the four categories of hedges are sufficient (in the sense that they embrace the hedges in my data), they are not disjoint. Bear in mind that in conjectural mathematics talk there is an affective subtext just below the surface of the propositional text. It is there because mathematics is a human activity: the participants care about the mathematics, but they also care about themselves, their feelings and those of their partners in conversation. In the next section I shall show how, on occasion, speakers use Approximators for Shield-like purposes. There is a good case, in fact, for speaking of shield*ing* and approximat*ing*, to emphasize the effect of hedges in the context of use, as opposed to the identification of some rigid lexical categories.

Particular Hedges in Mathematics Talk

The children whom I interviewed were aged 10 or 11, and were being invited to make mathematical predictions and generalizations. When they hedge, it is more often than not in order to implicate (in the Gricean sense) uncertainty of one kind of another. In other words, their hedges predominantly are, or have the same effect as, Plausibility Shields, deployed at significant and identifiable stages in the interviews. For the sake of maintaining coherence whilst sampling from the data, I shall examine when and how particular hedges, or small groups of hedges, are used.

'Maybe', 'think'

'Maybe' and 'think' are paradigm Plausibility Shields which can successfully convey a speaker's lack of full commitment to a proposition under consideration. It is necessary here to give more detail from Episode A of the Frances/Ishka interview, for immediate and future reference. I had asked the two girls to come to an agreement about the number of ways of making 10. Their discussion proceeds:

T6:12 *Frances:* There's one and nine.
   13 *Ishka:* Yeah.
   14 *Frances:* So that's one. Two and eight . . . and then there's
   15 *Frances and Ishka:* Three and seven.
   16 *Frances:* Four and six, five and five, six and . . . oh, that's the same.
   17 *Ishka:* Five ways?
   18 *Frances:* Maybe.
   19 *Ishka:* Mm, maybe . . . I think . . .
   20 *Frances:* What do you think?
   21 *Ishka:* We haven't had five five, have we?
   22 *Frances:* We have!
   23 *Ishka:* Oh OK, er . . .
   24 *Frances:* The others are, like, if you do six four, we've already done four six.
   25 *Ishka:* Mm [sighs].
   26 *Frances:* Shall we just say five ways?
   27 *Ishka:* There's about five.
   28 *Tim:* Er, I'd like you to be more convinced, Ishka. I mean, if it's about five, then it's four or six or seven or whatever . . . the number's sufficiently small that I think you should be sure one way or another.
   29 *Frances:* I think it's five ways.
   30 *Ishka:* But I'm sure.
   31 *Tim:* You are sure?
   32 *Frances:* Me too.

Having enumerated five ways, Frances begins to repeat herself (T6:16) – 'oh . . . that's the same'. Rather, she offers me (and Ishka) the first insight into what sameness means to her in this context. She has an implicit criterion, which surfaces when she withdraws 'six and . . .'. Ishka evidently shares or accepts the view that reversals will not count separately, and she asserts (T6:17) that there are five ways.

The fact that Ishka's claim is tentative is indicated by rising intonation ('Five ways?'), which transforms her statement (that there are five ways) into a question. This is one of a number of instances where statements are hedged with rising intonation. Such prosodic hedges (the linguistic term 'prosody' refers to variations in pitch, loudness, tempo and rhythm) are effectively Shields.

Frances (T6:18) perhaps echoes Ishka's uncertainty, or perhaps she may feel that Ishka's answer is offered prematurely, before she has exhausted all the pairs she can bring to mind. In any case, the pair now seem to have an understanding that it will be productive to assert their uncertainty, and reconsider the 'five ways' claim. Ishka effectively conveys this (T6:19) in the form of two Shields without a substantive proposition. Frances encourages her ('What do you think?') to articulate her position. This is typical of a number of instances of apparent teacher-like behaviour by Frances in Episode A, at which phase of the interview she projects herself as the dominant, more confident partner. However, having encouraged Ishka, she is impatient at Ishka's next contribution, which only suggests that Ishka has forgotten what has already been listed. Frances indicates (T6:24) that she is now satisfied that no further possibilities have been overlooked.[4] There follows (T6:26, 27) an apparent reversal of the earlier roles (T6:17, 18) of Ishka and Frances in relation to the claim that there are five ways. On each occasion one has sought agreement with a hesitant assertion that there are five ways; the other has given hedged assent – Frances with a Shield, Ishka with a Rounder. The second time (T6:26, 27), however, I inferred that Frances was fully committed to the claim, whereas Ishka was not – 'I'd like you to be more convinced, Ishka.' I had, after all, introduced the dialogue cited above with a clear request for common consent:

> T6:10   *Tim:*   I'd like you two to agree between you . . . Incidentally, we'll adjust that [microphone], Frances, so it's not quite so close. Right, um, I'd like you just to – yours is fine, Ishka – I'd like you two to agree between you how many different ways there are of doing that. Right? Two numbers that add to ten, and I'll just be quiet for a moment.

My repeated request for agreement is complied with by Frances and Ishka to a remarkable degree, certainly in comparison with most of the pairs I interviewed for Make Ten. Ishka is not yet, however, prepared to concede unqualified agreement (T6:27). The function of her chosen hedge, 'about', will be considered later in this chapter. In any case, she has successfully implicated the fragility of her commitment, borne out by the fact that I press her quite explicitly on the matter of being (more) convinced, urging that she 'should be sure'. Frances responds with an apparently hedged (but see my discussion of the ambiguity of 'I think' later) indication of where she stands. Ishka's response is unhedged, fully committed – but is it genuine, or have I blackmailed her into renouncing doubt in order to please me? After all, what I have demanded is not 'the answer' but for Ishka to be 'more convinced'. Of course, I really wanted both! I am at this stage of the interview encouraging the children to generate valid instances of a generalization-in-waiting. That there are five ways of making ten is such an instance. I readily accept Ishka's assurance that she is 'sure' without comment as to whether or not she is right.

'*Maybe*' is a modal form which seems to be user-friendly, in that it is favoured by the children in comparison with the apparently synonymous 'perhaps' and 'possibly', which occur not once, in the children's speech or mine, in the whole corpus. The following transcript data illustrates the appearance of 'maybe' within hedged predictive statements:

> T9:133   *Tim:*   Alright. um, supposing, we've done, we've done ten, twenty, thirty, sixteen. [. . .] I'd just like you to sort of say how many you think there would be, say for the number twenty-four. [. . .]
>
> 136   *Rebecca:*   [. . .] twenty-two? No, not twenty-two ways. Twelve ways? [. . .]

140	*Rebecca:*	Oh yeah, on the twenty there were more ways than the six-teen, so on twenty-four . . . must be more than the twenty, because that was less, because it was a lower number.
141	*Tim:*	Right, how many more?
142	*Rebecca:*	I'm not sure. [pause]
143	*Runi:*	[whispers] Eleven and twelve, [inaudible, presumably 'forty'] ways.
144	*Rebecca:*	Not forty, fourteen.
145	*Runi:*	Yeah, that's what I was going to say
146	*Tim:*	Let's just see. Runi thinks **maybe** fourteen ways, and I think you suggested twelve, Rebecca, yeah?
147	*Rebecca:*	Yeah.
148	*Tim:*	Um, what was your reason for suggesting twelve?
149	*Rebecca:*	Well, it was four off, than twenty and then twenty-two [?twenty . . . two] was two less than four, so you've twelve. Have twelve because, if you had, twenty had ten ways and twenty-four was four more than twenty, then **maybe** it would be twelve, because it's, um . . . half way in between.
150	*Tim:*	'cos it's half way in between. OK. And what do you think, Runi, are you saying it's four more, so it's four more ways?
151	*Runi:*	Yeah, that's what I was thinking of.

When Rebecca explains her reason for suggesting twelve (as I put it), she seems to be reasoning that what happened in the increase from sixteen to twenty might happen again with a further increase to twenty-four. But, I argue, she is signalling an awareness that she may be jumping to conclusions by hedging (T9:149) 'maybe it would be twelve'. It is an honest and straightforward expression of doubt as to the validity of the reasoning and the conclusion. By contrast, I double-hedge in (T9:146) 'Runi thinks maybe . . .' as a device to cast doubt on Runi's unhedged – and incorrect – contribution ('fourteen ways') which is beginning to take over from Rebecca's interrupted – but correct – train of thought. We are some way into the interview, this is the fifth example I've asked them to consider, and I'm getting impatient. My reaction to Runi's off-course prediction is to undermine it by attributing doubt where there may have been none. Thus my 'Runi thinks . . .' (T9:146) is intended to imply 'but that's only what Runi thinks'. Furthermore, 'Runi thinks **maybe** fourteen' was intended (now I think about it) to convey 'even though Runi said fourteen, she wasn't really sure about it, and you shouldn't be either'.

'Think' (usually 'I think') is, in some respects, a straightforward hedged performative (Lakoff, 1973, p. 490); it appears to be the most frequently deployed hedge in my transcripts. For example, in this extract, Alex rejects my prompt to list ways of making twelve, and goes straight for a prediction. When I appear to question it, she affirms, hedges, then revises.

T3:43	*Tim:*	Any ideas about how many ways there would be say for twelve? For twelve you could have twelve . . .
44	*Alex:*	[instantly] Six.
45	*Tim:*	Six ways?
46	*Alex:*	Yeah, **I think** so. Seven. Twelve add zero as well.

There is, however, a potential ambiguity (Stubbs, 1986) associated with this, and with other 'private' verbs such as 'believe', 'suppose' and so on. I would characterize the distinction as between epistemic and root meanings. Let me offer an example:

> I:6.1   You could use calculus to find the minimum, but I **think** that completing the square would be more elegant.

The ambiguity here concerns whether the 'parenthetical' clause (Lyons, 1977, p. 738) 'I think' is being used to implicate:

- an uncertainty (epistemic meaning) concerning the validity of the substantive statement (whether or not completing the square would be more elegant); or
- a firmly held position (root meaning) – that completing the square would indeed be more elegant – arrived at after consideration along with other tenable positions. That is, an assertion of what I judge to be the case.

In speech, the intended force may be made more evident by the location of stress in the utterance, i.e. 'I *think* that . . .' for the epistemic meaning as opposed to '*I* think that . . .' for the root.

The extracts which follow are from my interview with Anthony and Sam.

> T7:5   *Tim:*   . . . What I want you to do is to talk to each other and come to an agreement about how many different ways you can do it. OK? [. . .] And I'll just listen for a moment. How many different ways can you do that?
>
> 6   *Anthony:*   Er, let's have a **think** . . . Halves, um . . .

[end of first extract]

> T7:43   *Tim:*   Eleven. Is that all the ways, do you [*root stress*] **think**, or are there any more? [Long pause] What do you **think**, Sam? Do you **think** there's any more ways, or do you **think** that's all the possible ways of doing . . .
>
> 44   *Sam:*   There's more.
>
> 45   *Tim:*   You **think** there's more? OK. What would – give me an idea of what another one might be, or what it might look like.
>
> 46   *Anthony:*   What about . . . you can put quarters into ten parts and like that, can't you?
>
> 47   *Tim:*   Mm . . .
>
> 48   *Anthony:*   Well if we put them in about nine parts . . . if it, all the way, keep doing that, you might end up to number ten.
>
> 49   *Tim:*   Ah, right, um . . .
>
> 50   *Anthony:*   If you started at one.
>
> 51   *Tim:*   Yeah.
>
> 52   *Anthony:*   And I got a bit less, bit less, bit less, bit less, about, um, slowly get to number two. Keeps going round. Yeah? Would that work?

53	*Tim:*	Right, I **think** I get the idea. I mean, can you sort of get us started, and we'll try and **think** it through together? [Pause] Mm hm?
54	*Anthony:*	Let's have a **think**. [Pause]
55	*Tim:*	Did you say we were dividing it up into ten?
56	*Anthony:*	Yeah.
57	*Tim:*	Right. So, what could one number be?
58	*Anthony:*	You could have it into eighths as well.
59	*Tim:*	Uh huh.
60	*Anthony:*	I've heard of eighths, um . . .

The interview with these two boys was unique in failing to develop a scenario for combinatorial generalization. This was chiefly because Anthony proposed using fractions at a very early stage, and then seemed driven by some irresistible internal force to consider denominators outside the scope of his competence. Matters were not helped by the fact that Anthony had been diagnosed as having an aphasic language disorder; over the years he had become expert in manipulating adults in conversation, diverting them away from questions or topics which he was unable to understand or cope with. One could view such strategies as lack of awareness of the co-operative principles, in particular of the maxim of Relevance. An alternative interpretation would be that Anthony is expert at flouting that maxim for his own ends. Since Anthony was more assertive than Sam, and I wanted to allow the children to have a part in shaping the agenda (my contingent questioning), the discussion had a somewhat surreal quality at times.

In T7:6 and T7:54 Anthony is 'simply' stating his intention to engage with a problem (T7:6) or task (T7:54). Actually, on the evidence of the data, this is not at all typically childlike (to announce an intention to think). It is very much the mark of the confident adult who 'formulates' a conversation – that is to say, who describes some feature or aspect of the conversation within the conversation itself (Garfinkel and Sacks, 1970). It is the response of the person – a lecturer, for example, holding forth within their specialist field – to a question for which s/he does not have an instant answer. The announcement 'I shall have to think about that' has the effect of:

- flattering the one who asked the question: it suggests that the question is non-trivial, so that 'even I', the expert, will need to pause for thought before I answer;
- making space for the speaker to arrive at a response, either by recalling some information that is not at the forefront of her/his memory, or (more impressively) by the exercise of reason upon available information. In either case the thinking is, as it were, on display for public observation, or even public entertainment;
- implicating uncertainty without undue discomfort. It is not uncommon for the presenter of a mathematics lecture (or lesson at any level) to leave unfinished the detail of parts of calculations or arguments in her/his

prepared notes so that students may be sure to witness a public (if somewhat contrived) resolution of uncertainty.[5]

With Anthony, however, what came across to me (and was later supported by
information gleaned about his social strategies for coping with aphasia) was a
plausible and well used device for mimicking co-operative intellectual effort, with
the intention of retaining, or even gaining, the teacher's goodwill. Note that my first
'I think' in T7:53 is epistemic, whereas the second is a proposal that we do some
thinking. I asked Anthony's class teacher what she made of his 'let's have a think'.
She pointed out that successive teachers, concerned about his apparently erratic
train of thought (and the psychologist's suggestion that he experiences difficulty
in connecting ideas to form coherent thought-sequences), would repeatedly have
offered him advice using a formula such as 'Now think about it first, Anthony'.
Add to this the fact that this is mathematics and I am perceived as the teacher. The
close, almost individual, attention I was giving Anthony would be very familiar to
him, as he was accustomed to having special, one-to-one support in recognition of
his learning needs. It is likely that Anthony construes my questions as being of the
'testing' kind (Ainley, 1988), whilst I intend them to function as what Ainley calls
'directing questions', to provoke the boys into thought about a problem. Anthony is
likely to be skilled in eliciting clues from teachers as he navigates his way through
the fog towards a response to their testing questions, to which they already know
the answers. Anthony wants me to feed him some more clues. For example:

> T7:52   *Anthony:*   And I got a bit less, bit less, bit less, bit less, about, um, slowly
>   get to number two. Keeps going round. Yeah? Would that
>   work?

The problem is that (1) I was unaware, at the time of the interview, of his particular
manipulative skills, and (2) he doesn't know I'm playing an interviewer's game
called 'contingent questioning'! Far from pulling him back on the rails when he
wanders off, I tag along with him. Anthony seems to be manipulating the situation
in order to delay genuine intellectual engagement with the problem posed.

'About', 'around'

Channell observes that 'about' and 'around' appear to be interchangeable
Approximators, and that the first is more common in speech. I shall examine here
their use by three different children, in two extracts from the data. The first is with
Harry and Alan.

> T4:39   *Tim:*   So how many ways is it, Alan?
>   40   *Alan:*   Nine.
>   41   *Tim:*   Nine, right. [Pause] What if instead of saying two numbers
>   adding up to ten I said two numbers adding up to twenty?
>   42   *Harry:*   That would be **about**, yeah, I think . . . that would be eighteen.
>   43   *Alan:*   [simultaneous] . . . Eighteen ways. **About** eighteen, probably.

The second, from Frances/Ishka Episode B, includes use of 'around'. In fact the pragmatic analysis of 'about' which follows could be applied equally well to 'around', and could be illustrated from this extract and elsewhere in the corpus.

T6:129	*Frances:*	Fifty?
130	*Ishka:*	**About** fifty yeah.
131	*Tim:*	**About** fifty. Now are you saying **about** fifty, Ishka, because you're sort of playing safe or, I mean, do you really think it is fifty?
132	*Ishka:*	Well maybe not exactly, but it's **around** fifty basically? [. . .]
134	*Frances:*	Maybe **around** fifty.

In each case a prediction is being made – the number of ways of making 20 (T4:42, 43) and 100 (T6:130) – and each time the hedge is an Approximator (a Rounder, in fact) at the surface level. Channell (1994, p. 46) has found that respondents typically understand 'about *n*' to designate a range of possibilities, symmetrical about the exemplar number *n*. So the boys predict that the number of ways to make 20 is in the region of eighteen, maybe more, maybe less. I suggest, however, that the deep-level purpose and function of the hedge is Shielding against possible error in the cognitive basis of their prediction. This suggestion is supported by closer inspection of the data in context. Harry and Alan have already listed ways of making 10, and decided on nine positive integer possibilities, allowing reversals but not including zero as a summand. On being presented with the second problem (making 20) it was more common for children to list and count again, as Frances and Ishka do in Episode A.

Harry, however, is a confident boy.[6] He is a risk-taker, and goes straight for a prediction for making 20, avoiding the tedium of listing and counting. The basis of Harry's prediction seems to be proportional reasoning (doubling) – there are nine ways of making 10, so there are eighteen for 20. For fuller insight into Harry's thinking, his next contribution (following T4:43 above) is:

T4:44	*Harry:*	No, I think nineteen.
45	*Tim:*	Eighteen, nineteen?
46	*Harry:*	I should write that again. [Laughs]
47	*Alan:*	What's that?
48	*Harry:*	Up to twenty.
		[Harry begins a list $10 + 10$, $9 + 11$, $8 + 12$]

Later, and before the list is complete, he ventures:

76	*Harry:*	I think that'll be nineteen.

From the outset, then, Harry is uncertain as to whether the 'answer' to my question is 18 or 19. We just don't know how he arrives at these two possibilities. If his prediction is an extension of his experience of making 10, then (as already noted) doubling would produce Harry's first prediction. A more detailed awareness of the nature of his list of ways of making 10 (which I tried to prompt in the later episodes of some Make Ten interviews) could have led to the second prediction. The fact that he articulates it ('No, I think nineteen') is all the more remarkable because his first, incorrect prediction is confirmed by Alan, albeit with something less than total commitment (T4:43). It seems, then, that Harry may be entertaining these two different predictions from the moment I ask about making 20, and he seems (T4:42) to be testing out the first possibility, not just for my consideration (and possibly Alan's) but also for his own:

> T4:42   *Harry:*   That would be **about**, yeah, I think . . . that would be eighteen.

The effect of the initial hedging is to allow himself some space for further consideration, and to declare uncertainty in the assertion which completes the sentence. In the end he resorts to listing and counting, presumably since he lacks sufficient confidence in either of his predictions to choose between them when I ask him to do so ('Eighteen, nineteen?').

My conclusion is that the hedge 'about', although classified as an Approximator, is being used by Harry in T4:42 principally to assist the communication of his propositional attitude; in particular, to serve Shield-like ends. Harry's attitude to his prediction, and my interpretation of it, is reinforced by his use of the prototypic Shield 'I think'.

Ishka implicates the same attitude with 'about' in T6:130. My next turn in that conversation is evidence that I (as interviewer) suspected this intention:

> T6:131   *Tim:*   **About** fifty. Now are you saying **about** fifty, Ishka, because you're sort of playing safe or, I mean, do you really think it is fifty?

What I infer from Ishka's 'about' (T6:130) is not that she has approximated the actual number of ways to the 'round' number 50, rather that she is in possession of a generalization, a conjecture which would lead to exactly 50 as prediction. In fact, it is normal practice to use round numbers as vague numerical reference points (Channell, 1994, pp. 78 ff.); indeed, a round number on its own may serve as a Rounder (i.e. without a prefix like 'about' or 'approximately'), as in, for instance:

> I:6.2   A suit like that would cost you £300.

The fact that round numbers are normally chosen with numerical Rounders[7] is further evidence in support of my suggestion that Harry and Alan (T4:42, 43) are deploying 'about' as a Shield, and not as a Rounder. If their intention had been to approximate rather than to hedge commitment, then Channell's findings would lead me to expect 'about twenty' rather than 'about eighteen'.

My spoken contribution, then, in T6:131 is designed to test out Ishka's commitment to 50, asking, 'do you really think it is?' Again, my use of 'think' here is in the root sense of 'believe', and I strengthen the probe by the adverbial Adaptor hedge 'really'. The intended effect is to encourage her to make her position 'less fuzzy', as Lakoff puts it. Ishka's reply indicates her discomfort; she skilfully sidesteps my demand for commitment with a reply (T6:132) which amounts to a virtuoso performance in hedging.

132　*Ishka:*　Well, maybe not exactly, but it's around fifty basically? [. . .]

Even Frances, who at that stage is displaying more confidence than Ishka (and less hedging), double-hedges her response (T6:134).

'Basically'

This is an interesting and relatively unusual hedge, used by only three of the twenty-one children, and only on this one occasion (132) by Ishka.[8] The word can function as a 'bottom line' underpinning, synonymous with 'fundamentally, as in

I:6.3　John's problem is that he is basically lazy.

It seems to have the effect, as used by the children, of qualifying the content of what is being said or claimed; thus it acts as an Approximator. The following extract is from an earlier, weakly framed conversation with Simon (aged $12^3/4$), which turned out to be a forerunner of the Make Ten task. Simon rapidly moved on from positive integer pairs to decimals. After a while, I intervened:

Si1:15　*Tim:*　What if I gave you one of the numbers, one point three recurring, what would the other number be?
16　*Simon:*　Em, eight point six recurring.
17　*Tim:*　Why?
18　*Simon:*　Because one point three recurring is **basically** a third . . .
19　*Tim:*　You mean the point three . . .
20　*Simon:*　. . . point three recurring is **basically** a third, so you need . . . well, the one, that's one, so to make it up to nine you add on eight, then you need another two-thirds, which is point six recurring.
21　*Tim:*　If you have, um, point three and point six recurring, and you add them up, what do you get?
22　*Simon:*　Point nine recurring. Mmm – nearly one.
23　*Tim:*　Nearly one.
24　*Simon:*　Yes.
25　*Tim:*　Why nearly one?
26　*Simon:*　Because it's not, because point three isn't, it's just nearly a third. It doesn't quite get to the third.
27　*Tim:*　When it's point three recurring.
28　*Simon:*　Yeh.
29　*Tim:*　Oh, so point three recurring isn't really a third at all?
30　*Simon:*　Well, it's very nearly a third.
31　*Tim:*　Very nearly a third.
32　*Simon:*　Yeh.

Simon's statement (Si1:18, 20) that 'point three recurring is basically a third' is not in fact an assertion of a fundamental (basic, so to speak) property of point three recurring. The adverb 'basically' is being deployed as a hedge, a Rounder in fact, so that the force of the statement is much the same as that of 'point three recurring is approximately a third', or perhaps 'point three recurring is as good as a third'. It is as near as makes no difference. Brown and Levinson include 'basically' in a list of a dozen 'Quality hedges' (1987, p. 167), most but not all of which are archetypal Rounders ('approximately', 'roughly'), which 'give notice that not as much or not as precise information is provided as might be expected'.

Trace now the course of the above exchanges as the force of Simon's 'basic-ally' is revealed in the questioning. My analysis goes like this: as it stands, Si1:18 is 'incorrect' – not that the true/false dichotomy is very meaningful when applied to hedged assertions (Lakoff, 1972) – although the intention is clear to me. In Si1:20, Simon responds to my prompt to correct, or perhaps to clarify his statement in Si1:18. In fact, he interrupts my prompt to self-correct and (re)states that 'point three recurring is basically a third'. On the other hand, he completes the arithmetic in Si1:20 with 'you need another two-thirds, which is point six recurring'. Notice that there is no 'basically' this time. I (in my role as interviewer) am aware that confusion about the value of infinite decimals is commonplace with students – of all ages. This is not intended to be a patronizing remark, given the range of foundational issues which underpin any position on the matter (Cornu, 1991). The issue here is the usual psychological and notational difficulties associated with equating an infinite series (the decimal) with its sum (the fraction). My strategy, in order to ascertain where Simon stands in relation to these two recurring decimals – determined 'on the hoof', as the 'um' (Si1:21) indicates – is to ask him about their sum. As I expect, his reply conveys his belief that the sum falls short of one. Asked to explain, Simon is guarded but more explicit:

Si1:26 . . . because point three isn't, it's just nearly a third. It doesn't quite get to the third.

I press the conclusion in Si1:29, the 'Oh' attempting to convey some neutrality, some surprise, so as not to put words into his mouth. But he remains uncertain, and unable to agree without qualification to the bald statement that 'point three recur-ring isn't really a third at all'. His reluctance to concede is marked by the maxim hedge 'Well' (Si1:30) as he flouts the maxim of Manner, and arguably others besides. The whole exchange is marked by Simon's desire to be co-operative, yet true to himself, his beliefs and his uncertainties.

'Well': Maxim Hedges and Discourse Markers

Carlson (1984, pp. 37–38) analyses the occurrence of 'well' in dialogue in terms (*inter alia*) of failure to meet the demands of a question. Speakers tend to preface answers with a discourse particle such as 'well', to indicate some sort of insufficiency in the answer to be given (R. Lakoff, 1973). Examples given by Carlson (who makes an interesting choice of literature as data) include the following from Agatha Christie (1977, p. 22).

> Reflect a minute, Hastings. One can catch a murderer, yes. But how does one proceed to stop a murder? Well, you – you – well, I mean – if you knew beforehand – I paused rather feebly – for suddenly I saw the difficulties.

The following fragment is taken from one of my conversations with Simon (Rowland, 1997a):

Si0:1   *Tim:*   How many three-quarters are there in a hundred?
  2   *Simon:*   **Well**, there are seven three-quarters in ten, remainder a quarter.

In effect, 'well' acts as 'maxim hedge' in such instances (Brockway, 1981) – the speaker is serving notice to the hearer that the contribution about to come will in some respect fall short of the requirements of one or more of Grice's maxims. In these examples, adherence to the maxims of Manner (Christie) and Quantity (Simon) are in question. Perera (1990, pp. 217–222) found that 'well' occurs the most frequently of eight 'characteristically oral' constructions that she examined in the Fawcett corpus (Fawcett and Perkins, 1980) although she offers no pragmatic account for this observation. There is, however, quite a substantial body of literature on such particles (for example, Wierzbicka, 1976; Carlson, 1984; Schiffrin, 1987; in addition to the work of Lakoff and Brockway).

Wierzbicka analyses 'well' as a 'pragmatic particle', a word whose function is to express a pragmatic meaning at minimal cost. She uses the term 'pragmatic meaning' to refer to factors of propositional attitude such as assumptions, attitudes and intentions. Considered against the backdrop of Grice's maxims, 'well' can frequently be argued to attach some vagueness to the speaker's compliance with one or more of the maxims. Indeed, the device is not uncommon in the Make Ten interview data:

T10:181   *Tim:*   Thirty-nine ... why, how do you know that?
  182   *Susan:*   **Well**, you've got the, you've got your, let me see, nineteen ways, and then you've got another set of nineteen ways going the other way.

It could be argued that Susan can foresee rather a rambling account ahead, likely to violate the maxim of Manner. 'Self-repairs' (false, starts, self-corrections) abound in T10:182. This is commonplace (in the transcript data) when people are asked to supply a reason for a belief, or an explanation of some sort. In the example below, I am talking with Lucy and Rachel: I have introduced the conversation with my usual gambit, and whilst Lucy sets about listing ways of 'making' 10, Rachel quietly indicates that there will be ten ways.

T2:14	*Tim:*	So you're saying, Rachel, even before Lucy's written them all down, you're saying that there'll be ten ways.
15	*Rachel:*	Mm [in gentle confirmation].
16	*Tim:*	How did you know that, before Lucy had written them all down?
17	*Rachel:*	**Well**, because if you've got a number that adds up to ten, the, em, there's ten, and you've got all the others down below. You can only make ten ways to get up to ten.
18	*Tim:*	Do you understand that, Lucy? [Tim doesn't. Lucy nods] I don't. You explain it to me.
19	*Lucy:*	Em . . . **well** . . . there's only ten ways to make ten.
20	*Tim:*	**Well**, I can see you've only, I mean you've written down ten ways, right? If you count them up there's one, two, three . . .
21	*Lucy:*	. . . and they're all different . . .

In T2:16 I ask Rachel for an explanation: how did she know there would be ten ways before they were listed? The 'well' with which Rachel begins her explanation gives notice that I shouldn't expect an account which is entirely clear or convincing. Next Lucy is put on the rack, asked to clarify Rachel's argument. In fact she is only able to restate the conclusion, and her 'well' is encased in hesitation.

Brown and Levinson (1987, pp. 164–171) describe how some occurrences of hedges themselves (as opposed to discourse particles such as 'well', 'after all', 'anyway') may be interpreted as acting as maxim hedges. These hedges may do one of the following:

- emphasize that one or more maxim requirements is being met:

| T2:193 | *Roksana:* | **I do believe** there are thirty-eight now [maxim of Quality] |

- serve notice that (or indicate the possibility that) one or more maxim requirements is being flouted:

| T4:99 | *Tim:* | You think it might be seventy-two, Harry? |
| 100 | *Harry:* | It's a wild guess, but I . . . I'm not sure about that . . . I think it will probably be an even number . . . most likely about that, because . . . yeah, that is quite likely. |

Harry's highly equivocal response flouts the maxim of Manner. The hedges simultaneously achieve that effect and point to it.

A CA Perspective on 'Well': Adjacency Pairs

An alternative, or complementary, way of understanding the use of 'well' is in terms of preference organization with respect to adjacency pairs. In Chapter 4, I described how dispreferred second turns are marked in various ways, including delays and prefaces. The use of 'well' to preface a reply to a request for information in the classroom can certainly be interpreted in this way. As Levinson (1983, p. 334) remarks, 'The particle "well" standardly prefaces and marks dispreferreds'. Some uses of 'well' in the Make Ten corpus (and elsewhere in my data) can be seen to be of this kind. In the extract above with Lucy, my first part (T2:18) is an invitation (to explain); Lucy's second part (T2:19) amounts to a refusal, or at least inability, to accept the invitation.

My response (T6:20) indirectly evaluates Lucy's 'explanation'; in this case the indirectness marks my redressive action in anticipation of a Face-threatening Act.

There is additional evidence of the importance for pupils of 'well' as a pragmatic particle earlier in this chapter, in the Frances/Ishka Episode B (T6:118), and in Simon's utterance (Si1:30) towards the end of the previous section.

Hedges: the Taxonomy Revisited

The account of hedges in mathematical discourse in this chapter draws on the taxonomy of Ellen Prince and her collaborators (Prince et al., 1982), who identify two main categories: Shields and Approximators. These are illustrated by:

> I:6.4   I think there are ten beans in the jar. [Plausibility Shield]

> I:6.5   There are about ten beans in the jar. [Rounder-Approximator]

In each case, a hedge fuzzifies the sentence:

> I:6.6   There are ten beans in the jar.

The statement I:6.5 is arguably true if the number of beans in the jar is in fact eleven, whereas I:6.6 is not.[9] Thus, Approximators modify a sentence in a way that has truth-conditional semantic consequences. One view (Sadock, 1977, p. 434) is that Approximators not only alter the conditions under which a statement is true, but, by virtue of their vagueness, they 'trivialise' its semantics and so render it 'almost unfalsifiable'. (Note Sadock's own use of an Adaptor, making his *own* claim almost unfalsifiable.) This is a semantic observation. A pragmatic perspective on the same claim, taking into account goals and intentions, ought to consider whether the speaker intended it to have that effect, and if so, why.

In contrast to Approximators, Prince et al. claim that 'Shields [. . .] do not affect the truth conditions of the propositions associated with them' (1982, p. 89). I suggest that, in fact, the function of a Shield is to transform a proposition into a non-propositional speech act. The syntactic means of doing this is the use of a hedged performative ('I think'), an adverbial preface ('probably', 'apparently') or some other fuzzy meta-linguistic device (see Stubbs, 1986 for others). The illocutionary force of such constructions is that what was an (unhedged) statement sheds its status as a proposition (subject to truth-conditional semantics) to become a conjecture. The crucial effect is that the speaker has less stake (or none) in the truth or falsity of the (unhedged) statement.

For example, suppose I say:

> I:6.4   I think there are ten beans in the jar.

I then carefully count the beans in the jar, and find that there are indeed ten. This would be sufficient (irrespective of what I might believe about the number of beans in the jar) to render the statement I:6.6 true; but it does not render I:6.4 either true or false, since the conditions under which I:6.4 are true are strictly independent of the number of beans in the jar. They have to do with my beliefs, my propositional attitude, my state of mind. The pragmatic effect of I:6.4 is to imply that the speaker doesn't know exactly how many beans there are (since otherwise s/he would be violating the maxim of Quantity), that s/he entertains the possibility that there are ten, but is unwilling to be held to be committed to the truth of such an assertion.

Consider once more:

I:6.5   There are about ten beans in the jar.

Now Approximators are deployed (at times) precisely for the purpose of constructing a scenario within which a proposition is (almost) unfalsifiable. Why should a speaker want to do that? One reason would be that s/he is uncertain and does not wish to be seen to be committed to a straightforward proposition for fear of being seen to be wrong. As with the Shield, the illocutionary force of the statement is conjectural. The suggestion of Prince et al. that (p. 95) 'Rounders do not reflect any uncertainty or fuzziness but are rather a shorthand device when exact figures are not relevant or available' is somewhat hasty; indeed, it is ultimately untenable.

There are, in effect, two parallel taxonomies of hedges:

- Into lexical categories as specified by Prince et al. Thus 'about' is (i.e. has the typical form of) a Rounder, and so on.
- Into the following two semantic categories, in parallel with that of modal verbs (Coates, 1983, pp. 18–22):

  Epistemic hedges – conveying a state of mind such as lack of knowledge, (un)certainty, (lack of) conviction, commitment, etc.
  Root hedges (after Coates, 1983, see below) – non-epistemic: for example, giving an appropriate degree of precision. A variety of pragmatic purposes, including those listed by Channell (1994), may motivate such a hedge, but displacement is not one of them.

The essence of my argument in this section is exemplified by what I see as the two different pragmatic meanings of the Approximator 'about'. The first is epistemic, and occurs in my data in situations requiring prediction or generalization, such as

T4:43   *Alan:*   Eighteen ways. **About** eighteen, probably.

The pragmatic goal of 'about' is to implicate uncertainty, and to achieve protection against an accusation of error by rendering the utterance unfalsifiable.

The second is root, normally (but not necessarily) associated with estimation. In this dialogue, C is a confident 10-year-old girl, I an adult interviewer.

M222:1	*I:*	Can you tell me how many sweets there are on the plate?
2	*C:*	**About** twenty?
3	*I:*	Now, can you tell me how many sweets there are in the glass?
4	*C:*	Ten.
5	*I:*	And do you think there are exactly ten?
6	*C:*	No! [laughs] **Not exactly.**

No counting was involved. The answers (M222:2, 4) are rapid estimates; her amusement in M222:6 makes that clear. The Approximator 'about' is implicit in M222:4, ten being a 'round' number (Channell, 1994, pp. 87–89). Here the pragmatic goal is to meet co-operative requirements to do with Quality and Quantity – not to make claims in excess of one's actual knowledge, and to judge how much detail is required in a given situation. These matters will be considered again in the next chapter.

In pupils' mathematics talk associated with predictions and generalizations, the semantic function of Approximators is usually epistemic. Similarly (but not so frequently), the semantic function of a Shield may be root, e.g. the 'private' verbs 'think', 'believe', and so on.

The epistemic/root approach focuses attention on the function of hedges irrespective of their form: it also facilitates discussion of properties of hedges that have a great deal in common with those of modal verbs.

Hedges and Politeness

The use of hedges can often be seen as a means of redressing threats to 'face'. Consider, for example, the utterance:

I:6.7   I'm sort of hoping to get it finished by Friday

in which the speaker's commitment actually to finishing by Friday is loose, and there is only the weakest sense of any kind of promise (a threat to negative face). This is very typical of the use of epistemic hedges in Make Ten.

T9:149   *Rebecca:*   ... maybe it would be twelve, because it's, um ... half way
           in between.

The epistemic modal 'would' and the epistemic hedge 'maybe' redress the potential face threat to Rebecca (in case it turns out not to be twelve).

This is broadly consistent with my observation that I, as interviewer, typically use Shields and Adaptors in recognition of the face wants of the children, whereas they typically use Rounders and Plausibility Shields as epistemic hedges which render their conjectures almost unfalsifiable, in order to serve their own face wants.

Students who, in their own perception, enjoy a more balanced power relationship with their tutor do in fact have concern for his or her face needs. This was apparent in a small way with the 10- and 11-year-olds when they felt they might be usurping my role as 'teacher', and exhibited negative politeness. Here, for example, Caroline would like to explain something:

T3:53	*Caroline:*	Um, can I . . . ?
54	*Tim:*	Oh yes, please, Caroline.
55	*Caroline:*	Half of ten is five and we actually got five, but then we added on ten add zero, so it would be six. So, so far it's basically worked out half the number, and it's the same with the twenty, it's eleven because we've added on zero and twenty, and it's the same with the sixteen.

Caroline adopts the face-redress strategy of posing her offer in the form of a question, before giving an extended explanation. Note the relative absence of hedging in T3:55 – here she is not coming to know the matter she articulates; rather, she *knows* it.

The Zone of Conjectural Neutrality

In this chapter I have shown how children use Rounders and Plausibility Shields to implicate uncertainty, to insert some space between conviction and asserting a proposition. I have found it useful to give a name – the 'Zone of Conjectural Neutrality' (ZCN)[10] – to the space between what we believe and what we are willing to assert. Even Rounders, such as 'about', which syntactically attach some fuzziness to the proposition itself, are pragmatically deployed by the children to achieve Shield-like ends. This, and the forms of linguistic Shielding which I have discussed, have the effect of reifying the ZCN and thus distancing the speaker from the assertion that he or she makes. Whilst truth and falsity may be decided in the ZCN, a person may articulate a proposition without necessarily being committed to its truth. In such a cognitive and affective *milieu* it is the proposition that is on trial, not the person. Whilst mathematical conjectures are formed as private, cognitive (perhaps inductive) acts, they are validated in public polemic of some kind. Moreover, the learner ideally participates in the discourse, since, as Balacheff submits (1990, p. 259), children must take responsibility for the validity of their own solutions 'in order to allow the construction of meaning'. At the same time, a conjecture is not fixed and immutable, but modifiable. I am describing, of course, the quasi-empiricist approach to teaching and learning. In Chapter 3, I referred to Dawson's (1991) account of a 'fallibilistic way of teaching'.

> A teacher who is functioning fallibilistically [. . .] establishes a classroom climate in which an atmosphere of guessing and testing prevails, where the guesses are subjected to severe testing *on a cognitive rather than an affective level* [. . .] where knowledge is treated as being provisional. Because of the provisional nature of knowledge, pupils are encouraged to confront the mathematics, their peer group and, where appropriate mathematically, even their teacher. (Dawson, 1991, p. 197, emphasis added)

Not only is uncertainty an intellectually tenable position, but the assertion of uncertainty draws the attention of the teacher to the existence of a ZCN, and thus opens up the possibility that s/he might provide the student with some cognitive scaffolding (Wood et al., 1976) to support, and perhaps transform, that state. This seems to be what is happening to Harry here, in a final extract:

> T4:128   *Tim:*   How do you know there are forty-nine, Harry?
>
> 129   *Harry:*   Well, **I am not completely certain, actually**, but I would expect it, because if you start off with fifty and you do forty-nine add one, forty-eight add one, but then you'd end up with one add forty-eight, wouldn't you, so they always change . . . [. . .]
>
> 133   *Harry:*   Forty-eight add two, I mean.
>
> 134   *Tim:*   OK. And the last one in that list would be?
>
> 135   *Harry:*   One add forty-nine, so they'd all be . . . [interrupted by Alan sneezing]
>
> 136   *Tim:*   How do you know that there's forty-nine different ways that you've listed? You started with forty-nine add one and you ended up with one add forty-nine. Now how do you know that there are forty-nine pairs in that list?
>
> 137   *Harry:*   Well, there's fifty numbers, and you just, there's lots of ways because you just go forty-nine add one, forty-eight add two all the way down till you get to the one, but you can't do fifty add nought, so that will take away one which will leave you with forty-nine. **I'm quite certain about that.**

I shall return to further consideration of the ZCN in the final chapter.

Summary

In this chapter, I have shown that the classification of hedges (due to Prince et al., 1982) into functional categories is relevant and useful in the analysis of my task-based mathematical conversations with children aged 9 to 12, where children are being encouraged to predict and generalize. I have noted that:

- I (as interviewer) exploit Attribution Shields and Adaptors, usually for teacher-like purposes; *whereas*
- the children typically use Rounders and Plausibility Shields, and nearly always to imply uncertainty, to insert some space between conviction and asserting a proposition. *Furthermore,*
- I have proposed that the space between what we believe and what we are willing to assert should be recognized, and that it should be named the 'Zone of Conjectural Neutrality'.

The purposes which speakers achieve by the use of vague expressions (Channell, 1994, pp. 186–189) include 'displacement' (in case of uncertainty) and 'self-protection' (a safeguard against later being shown to be wrong). Given the prevailing school culture (maths is about right and wrong answers, and it is much better to be right), the use of hedging is evidently deployed by many children as a Shield against being 'wrong'. These Shields could be seen to act as linguistic pointers to intellectual risks, with attendant vulnerability. In principle, it would be preferable for students to know that being unsure is a genuine, valuable and creative option available to them.

Notes

1   Pimm (1987, p. 52) observes that 'transcripts of actual classes regularly indicate little verbal interaction between pupils themselves (particularly about mathematics)'. I had in mind that this is frequently the case; this is not to accept that it has to be the case, only to recognize that it was not my aim (in this research) to attempt to change classroom culture.

2   I intend the word 'gambit', as I use it in this chapter, to mean little more than an opening move. In relation to teacher questioning, Pimm (1987) associates the same word with the possibility of sacrifice. In my case, this is appropriate to the extent that, in allowing the children some measure of control (over the interpretation of the offered task), I may lose control of the direction of the interview in terms of engagement with certain mathematical processes.

3   The mathematical consequences of such choices may be summarized as follows. Let n be a positive integer and $f(n)$ be the number of pairs (a, b) such that $a + b = n$, where a, b belong to a set A of 'numbers'. If (b, a) is taken to be distinct from (a, b) (unless $a = b$) and A is the set $N = \{1, 2, 3, \dots\}$ of natural numbers, then $f(n) = n - 1$; if A also includes zero then $f(n) = n + 1$. If, however, (a, b) is always identified with (b, a), and $A = N$, then $f(n) = \frac{1}{2}n$ when n is even, and $\frac{1}{2}(n - 1)$ when n is odd. With zero included in A these become $\frac{1}{2}n+1$ and $\frac{1}{2}(n + 1)$ respectively. Of course, if A includes the set of integers, then $f(n)$ is not finite.

4   The use of a linguistic formula such as 'like, if you do' to refer to a general relation or a general process – in this instance additive commutativity, or symbolic reversal – by means of an instance of that relation/process is commonplace. It is another instance of the power of the generic example to evoke well founded confidence in a related generality. See also the discussion of 'you' in Chapter 5.

5   The first two of these three points relate to Wierzbicka's analysis of 'well' as a 'hesitation noise' (1976, p. 330).

6   Harry had recently transferred to the state-maintained school from a famous independent preparatory (private elementary) school. Transfer documents from the prep. school gave little indication of Harry's actual attainment, but did observe that 'Harry is the only boy in his form who has not obtained his own copy of the *Odyssey*'. This remark could be interpreted as a revelation of Harry's independence of thought and action, and of his willingness to take risks. In any case, such comments assist those who inhabit different cultures (including the majority of English people) to understand what really matters in the great English private schools. Harry's apparent negligence with regard to the classics, unlike other manifestations of his independent spirit (or indolence), proved not to disadvantage him at his new school.

7   But not invariably. Here (noted December 1994) Mark is asking his mother, Judy, about a Christmas present for his grandfather:

> *Mark:*   How much do you think a Ruth Rendell book will be?
> *Judy:*   About four ninety-nine.

8   Children seem to latch on to their preferred hedges. Whilst 'basically' is rare in my data, see, for example, Maher et al. (1994, pp. 213–214), where the use of 'basically' by the boy Alan is striking, and very much consistent with the analysis I give for Simon.

9   A strict truth-conditional interpretation of the sentence I:6.6 would be to say that it constitutes a statement which is true provided the number of beans in the jar is at least ten. A standard pragmatic view, however (Levinson, 1983, p. 106) is that a person uttering I:6.6 implies 'ten and no more' because the hearer expects adherence to the maxim of Quantity. In everyday discourse, the pragmatic interpretation is assumed; indeed, I would suggest that the truth-conditional interpretation would be considered rude or ostentatiously 'clever'. There is some analogy with the truth-conditional interpretation of the question 'Would you like tea or coffee?' which admits the answer 'Yes'.

10 See also Rowland (1997b). My choice of the name 'Zone of Conjectural Neutrality' for this space between articulation and belief was prompted by Vygotsky's term 'zone of proximal development' for the gap between what a learner can do alone and what s/he can achieve with 'expert' assistance. The two zones are different, however. Hewitt (1997) independently gives the name 'neutral zone' to a region which contains the collection of sensory stimuli ('offerings') from which a student selects a subset for her/his attention. Whilst these are three different kinds of 'zones', it is not difficult to see meaningful relationships between them.

7   Estimation and Uncertainty

Tip No. 448: Don't be afraid to say, 'I don't know.' (H. Jackson Brown, Jnr, *Life's Little Instruction Book*, 1991)

From time to time a pupil may feel obliged to make an assertion, perhaps in answer to a question, yet without certainty that what s/he is claiming is entirely accurate or true. The 10- and 11-year-old children considered in the previous chapter had available a repertoire of hedging strategies for maintaining co-operative interaction whilst being appropriately vague, thereby conveying a lack of full commitment to the propositional content of their utterances. When is this linguistic repertoire developed, and are there identifiable milestones on the way to confident mastery?

The vague language identified and discussed in earlier chapters arose in contexts where pupils were engaged in activities involving prediction and generalization. This required extended, contingent interviews, in order to prepare the ground, i.e. the problem environment, for these mathematical processes to come into play. Much of Channell's book (1994) on the pragmatics of vague language is concerned with approximating *quantities*. The study reported in this chapter focuses on that dimension of vague language, specifically on *estimation* of the number of objects in a set. This choice of focus is partly for the sake of addressing what is perhaps the most obvious aspect of mathematical activity in which one would expect vague language to play a part. Moreover, it is possible in a short (5–10 minute) interview to present appropriate estimation tasks to children in a meaningful way, to obtain responses, and to follow these up from a restricted menu of probes. It is therefore convenient, in designing an age-related study, to use estimation rather than generalization tasks to elicit vague language when dealing with a pupil sample numbered in hundreds rather than tens.

Estimation

Clayton (1992, p. 11) classifies the diffuse notion of estimation into three broad categories.

Computational estimation involves the determination of approximate (typically, mental) answers to arithmetic calculations, e.g. 97π is roughly 100×3.1 or 310. Such competence is commended by the National Curriculum for Mathematics in England and Wales (DFE, 1995, p. 25) for the purpose of checking answers to precise calculations for their 'reasonableness'; pupils, however, seem to regard such checks as trivial or pointless (Clayton, 1992, p. 163).

Quantitative estimation indicates the magnitude of some continuous physical measure such as the weight of a book, the length of a stick.

Numerical estimation entails a judgement of 'numerosity' – the number of objects in a collection. In principle, such a set could be precisely quantified by counting. In practice such precise enumeration may be impracticable or simply judged to be unnecessary, excess to pragmatic requirements.

Ellis (1968, p. 159) observes that counting may be considered to be a measuring procedure, unique in the non-arbitrariness of the unit of measure. In fact, Clayton merges numerical estimation and quantitative estimation into one analytical category. Sowder (1992) notes that 'there simply is not a rich research base in estimation' and that most such research has been on computational estimation. Moreover, 'Numerosity estimation has received the least research attention, and [. . .] the only two studies located combine it with measurement estimation' (p. 372). One those two studies was reported in a short article – by Clayton himself – in *Mathematics Teaching* (Clayton, 1988). The other study (Siegel et al., 1982) analyses numerosity estimation competence in terms of 'benchmarks' (known standards) and 'decomposition/recomposition' of a set in order to apply a benchmark together with a computation. A variety of estimation tasks (of quantity and numerosity) were presented to children aged between 7 and 14, and to a small sample of adults. The investigators found marked developmental differences in performance on numerosity items. The number-estimation tasks (such as 'How many names on a page of the phone book?') seemed to involve quite large sets.

The literature on estimation of quantities does indeed tend to be about the estimation of measures. Ainley (1991), setting out to investigate the *mathematics* in measurement, describes a staple-diet 'estimate then measure' lesson with 8-year-olds. Reflecting on the lesson, Ainley comments (p. 70) on the peculiarity of estimating and *then* measuring. Clayton (1988, p. 23, p. 158) independently agrees:

> Most estimation tasks in school require an estimate and then (almost immediately) a measure or calculation is made [. . .] pupils often make their 'estimate' **after** they have measured or calculated, showing their disregard for the estimation process. (p. 23)

The behaviour of the children in Ainley's account supports Clayton's observation; some of them enter the measurement performed by another child for their 'estimate' – it is, after all, so much more satisfactory if the two agree. Ainley concludes that:

> There is mathematics in measurement; but it does not happen to be in the bits which currently get given priority in mathematics lessons. (p. 76)

The study to be presented in this chapter is concerned with children's estimates of numerosity, and their spontaneous production of vague language in articulating and discussing such estimates. Suppose, for example, that I ask a 'phone book' question such as 'How many words are there on this page?' You (the reader) are likely to respond – perhaps from experience of reading essays or writing papers – along the lines 'About four hundred'. The hedged approximation in such a response can be accounted for by reference to one or more of the pragmatic goals listed towards the end of Chapter 3.

The first of these goals (giving the right amount of information) accords with Grice's maxim of Quantity – 'Let your contribution be sufficiently informative but not excessively so' for co-operative interaction. The point is that, in saying or writing 'About four hundred', you have judged that I don't much care whether there are actually 388 or 413 words on the page. The fourth goal (covering for lack of specific information) could also account for the same hedged response. You don't know precisely how many words there are, and can't be bothered to count them. At the same time, by saying 'About four hundred' you do observe the maxim of Quality – 'Be truthful: don't say that for which you lack evidence'.

The last goal in Channell's list – protecting oneself against making mistakes, against accusation of error – is equally pragmatically plausible. For one consequence of the vagueness of the response is that it would be very difficult to demonstrate that the claim contained in it was wrong. The hedge is epistemic, and works for the speaker because it effectively renders the claim unfalsifiable.

Prince et al. (1982) and Channell (1985, 1990) have demonstrated how speakers and writers deploy hedges in order to fulfil a variety of goals. Their studies were mainly of adult academic and professional groups, such as doctors, copywriters, broadcasters, students of linguistics and economists. Both identify the prevalence of a protective purpose – the recognition of vulnerability and the consequent need to need to 'shield' oneself. This need is ever-present in those public settings, usually schools, where pupils do mathematics.

Ainley (1991, p. 70) says, *en passant*, of the estimation lesson she observed: 'It is a relaxed lesson: estimates are meant to be wrong, so no one is worried about failure.' Perhaps the lesson is relaxed in the sense that it demands little of the children. But I wonder whether these and other children have been let in on the secret that estimates are expected to be wrong! If too many of a child's estimates agree with the 'right' (i.e. measured) answers, the teacher suspects foul play – a classroom form of Tiegen's paradox (Chapter 3) in that the child who is most accurate is deemed to be the least likely to have arrived at his 'estimate' by fair means. This is a strange mathematics classroom game, in which 'right' estimates are more likely to meet with the teacher's disapproval than plausible 'wrong' ones. That this is not known to the children is evidenced by the way many of them subvert the activity – estimate *after* measuring, not before. It is a first step on the ladder of innocent deceit: some rungs higher up, we find the science student who draws a straight line graph, then plots some points plausibly arranged either side of it, before finally tabulating his 'experimental' results. Notwithstanding a relaxed atmosphere in the classroom, I suggest that 8-year olds *are* worried about failure when they do estimate-and-measure in school. Not worried in a debilitating sort of way, perhaps, but enough to want to fix the answers.

Weiner (1972) identified a vicious circle in children's attitudes to estimation: poor estimators, not surprisingly, viewed estimation as 'risky', avoided it, and remained poor at estimation. Clayton (1992) presents and studies estimation as a risk-taking activity. Transcripts of pupils performing estimation tasks were analysed to determine a (somewhat subjective) measure on a scale, 1 to 10, of pupil confidence in the estimates they gave, using indicators such as 'willingness to explain methods, general air of confidence'. The judgement of confidence was irrespective of the suitability or accuracy of the estimate. Clayton's conclusion is that the boys in his sample were generally more confident than the girls. I return to this issue towards the end of this chapter.

Counting

The earlier question about the number of words on a page invites an estimate of numerosity – the cardinality of a discrete, finite set which could be counted, if an exact answer were thought necessary. One of the tasks presented to children aged 4 to 11 in the present study was designed to offer the choice of counting or estimating a set of nineteen objects. Data on the methods used by those children who did opt to count were collected and coded in the course of the study. These counting data are a kind of by-product of the design, but they turn out to have some relevance in interpreting the linguistic behaviour of the children who are presented with a mathematical task involving numerosity.

In the early years of schooling, children are explicitly taught to count, but not to estimate, small finite sets. In the UK there is currently an influential and persuasive movement (Thompson, 1997) urging teachers to place counting at the heart of arithmetic teaching in the early years, instead of sets, sorting and one-to-one correspondence. The process of counting has been minutely analysed, with detailed and illuminating studies by Zaslavsky (1973), Gelman and Gallistel (1978), Steffe et al. (1983), Steffe and Cobb (1988) and Fuson (1988, 1991). In essence, counting a finite set entails the matching of the elements of the set (in any order, but without repetition or omission) with a fixed set of words – number-names ('tags') – which must be produced for word–object matching in a canonical sequence. Gelman and Gallistel (1978, pp. 77–82) identify five organizing principles in young children's counting:

- the stable order principle – the tags must be drawn from a stably ordered list;
- the one–one principle – every item in a set must be assigned a unique tag;
- the cardinal principle – the last tag used is the cardinality of the set;
- the abstraction principle – the above principles can be applied to any collection;
- the order-irrelevance principle – the order of enumeration does not affect the outcome of the count.

Gelman and Gallistel conclude (p. 130) that the first three 'how to count' principles are learned in that order; for example, a child can reliably recite the list of number-names before s/he can assign them injectively to a set of objects. Fuson questions the invariability of this learning sequence, finding that it depends on the size of the set to be counted. In particular, for sets of cardinality above 16, it is the one–one assignment that causes most difficulty, and (not surprisingly) this is especially the case when the set is disorganized rather than being presented in a row.

Fuson also draws attention to the fact that, when people perform a count, they have to find some way of co-ordinating the word–object correspondence, and that this is achieved by two simultaneous kinds of pointing 'actions', or 'indicating acts'. First, the person doing the counting has to point systematically to the *objects* in some (complete, non-redundant) order. Second, and simultaneously, they must point (in the sense of drawing attention, at the very least their own attention) to the number-*word* which is to be matched with the objects as they point to them in turn. Fuson calls these two independent indicating acts 'local correspondences'. Both must be one–one if the count is to succeed – 'one word must correspond to one indicating act and one indicating act must correspond to one object' (Fuson, 1991, p. 31). The form of each of these indicating acts undergoes change with growing maturity, but invariably begins from an externalized paradigm – the earliest object-indicating act is achieved by touch: the word-indicating act is speech, specifically by counting aloud. The first is spatial-tactile, the second linguistic. It is as though nature had been careful to assign one task to each cerebral hemisphere.[1] Between the ages of about 3 and 6, and beyond, each of the two pointing actions attenuate to internalized versions:

> Both action parts of counting immovable objects – pointing and saying number words – undergo progressive internalization with age. Pointing may move from touching to pointing near objects to pointing from a distance to pointing from a distance to using eye fixation. Saying number words moves from saying audible words to making readable lip movements to making abbreviated and unread-able lip movements to silent mental production of number words. (Fuson, 1988, pp. 85–86)

In the study to be reported later, the whole gamut of these 'actions', the human repertoire of word–act–object associations, from touch–say to gaze–mute, was demon-strated in the data collected from 230 short interviews with children aged 4 to 11.

Enquiry Focus

When a teacher (or textbook, workcard, etc.) asks, 'How many?' young children (in the first two or three years at school) typically receive the question *as an invitation to count* rather than to estimate.[2] As Nunes observes:

> Language is clearly connected to many action schemas: for example, when you ask children 'How many . . . ?', they count. (1996, p. 74)

In the context of primary school mathematics the suggestion is a very plausible one. The teacher's numerosity question activates the counting action-response. Counting a small set may also be regarded as a less risky enterprise than estimating its cardinality. If such is the case the young child's answer-response to such a question is less likely to be modalized or hedged than the older child's. Furthermore, the young child will not be able to hedge until s/he has learned how to achieve that effect with language. My expectation, then, is that modal forms and hedges will be relatively absent in mathematics talk in early childhood, and that one may discern progressive development of modal/hedging capability and use in individuals through the years of primary schooling.

The aim of the enquiry was to examine the validity of this expectation, looking for trends in the responses of pupils across the 4 to 11 age range, in the context of cardinal estimation activity.

I have divided both modal forms (Chapter 3) and hedges (Chapter 6) into two broad categories, labelled (in both cases) epistemic and root. The epistemic category contains instances of language use (modals or hedges) which encode and serve to convey the speaker's attitude to or confidence in what s/he is saying. The category is therefore pragmatically determined but, particularly for modals, there are semantic parallels, i.e. what kind of modality (wish, conjecture, etc.) is it? The root category is, by definition, non-epistemic. A working test for a root modal or hedge might be that the speaker has little or no affective 'stake' in what s/he says. Thus deontic modals (of obligation and permission) would normally be root, as would an Approximator hedge that could be claimed to be motivated by concern to respect the maxim of Quantity.

In this study, both modals and hedges will be identified in the first instance by reference to their form. Hedges will be identified as Approximators or Rounders, and modality identified by the presence of modal auxiliaries. This is easily justified, since modal verb moods and tenses (see Chapter 3) are marginal in modern English (Stephany, 1986, p. 385), especially in speech, and so the modalizing function falls on modal verbs (auxiliaries).

In fact, there is little point in distinguishing between epistemic modals and hedges. Adverbial modal forms such as 'possibly' and 'maybe' are in any case included as hedges (Shields). Conversely, Stubbs (1986, pp. 18–19) takes the Shields 'I think/believe/guess/etc./ that' and 'It seems that' to be modal when they release speakers from total commitment to propositions (i.e. when, in my terms, they are epistemic hedges) as opposed to root uses, when they are used to make statements about what Stubbs calls 'private psychological states' such as dogmatic conviction.

Method

The study was carried out in a 4–11 primary school. There were some 230 children on the school roll. Every child was asked the same three 'How many?' questions in private, one-to-one interview. Details of the questions and related tasks are given below. The object was to test the expectation that the language of modality and hedging will be more commonplace among the oldest children (10–11) than the youngest (4–5), with some sort of continuum evident between these extremes.

The fieldwork was carried out by a research assistant (the 'interviewer') in the last month of the school year. The tasks had been piloted in another school so as to train the interviewer and refine the precise wording of the questions themselves. In their final form these were as follows.

- *Task 1.* The interviewer produces a white plate on which nineteen sweets have been placed so that each is visible. The sweets are similar in size and appearance to Smarties.[3] The child is asked, 'Can you tell me how many sweets there are on the plate?'[4]
- *Task 2.* The interviewer produces a high-quality colour photograph of a small glass containing fourteen sweets. These almost reach the rim of the glass. The child is asked, 'Can you tell me how many sweets there are in the glass?'
- *Task 3.* The interviewer shows the child two thin plastic tubes (both are about 25 mm in diameter and 100 mm high). One contains ten sweets, the other twenty. The interviewer says, 'There are ten sweets in this tube [indicates]. I know that, because I counted them when I put them in. Can you tell me how many sweets there are in this [indicates the other] tube?'

The materials for each task were produced in turn by the interviewer as they were needed. Why choose these particular three tasks? The point about the first is that the child can actually count the sweets if s/he chooses to do so, but may also make a reasonable estimate if s/he so chooses. The actual number is a determinate and accessible quantity; the child must decide whether it is required, and be aware that estimation is an option. I had considered some numerical variation for the youngest (age 4–5) children, placing just nine, or even six, sweets on the plate. At the same time, I had a preference for keeping the task constant across all the age groups, otherwise it would have been possible to account for differential responses by the fact that they had been given different tasks. I was also concerned that if a child could 'see' how many sweets there were by direct perception and 'subitizing' (Jensen et al., 1950), the issue of estimating would not arise as an option. In fact, there are good grounds for believing that the enumeration by counting of nineteen sweets would be an accessible option for the 4- and 5-year-olds in the sample. Whether the count was *accurate* was, for the purpose of the study, immaterial. Gelman and Gallistel's study reassures that:

> 4- and 5-year-olds can [assign tags to items] for set sizes up to 19 [. . .] young [meaning 3-year-old] children do *not* treat set sizes in excess of 5 as undifferentiated *beaucoups*. (1978, p. 111)

In fact, both the pilot and the survey proper vindicated not giving the youngest children an 'easier' task.

The second task was designed so that the precise number of sweets in the glass was indeterminate. It cannot reliably be determined by counting, since not all the sweets are visible in the photograph. Some kind of estimate is therefore necessary, and some degree of uncertainty is likely to be present in the situation, although the estimate may be guided by a count of the sweets visible in the photograph. Uncertainty may be lessened by naive interpretation of the two-dimensional image, i.e. failure to realize that some sweets which were present in the glass are not part of the photograph.

Similarly, in the third task, the precise number of sweets cannot be determined by counting, since not all are visible on the outside of the tube. A handful of the 230 children tipped out the contents of the tube and counted them that way! However, the height of the sweets in the second tube, relative to the first, is a possible guide to the number in it, given the fact that there are ten in the first. In that case, then, estimation may be guided by an elementary form of proportional reasoning, namely doubling (Hart, 1979, p. 99). An alternative perspective on this strategy would be to view it as 'regular decomposition/recomposition' (Siegel et al., 1982, p. 213). The contents of the second tube are decomposed into samples, each estimated to be the size of the given 'benchmark', i.e. the contents of the first tube; the two samples are then recomposed and the answer computed. The interviewer sought to understand whether any such strategy and inference was a factor, using probes such as 'How did you know that?'.

Tasks 2 and 3 were essentially intended to 'block' (Laborde, 1979, pp. 33–34) the possibility of complete solution by counting, in order to introduce an element of uncertainty and the need to estimate. All three tasks bore some superficial similarity to the numerosity tasks in the study of Siegel and his collaborators: their stimuli were all physical props or photographs, and their questions were of the form 'How many X's are there in/on this Y?'. The essential difference is that in their study the problems were presented as estimation tasks. The children were told that they were to be asked to make estimates, and were given a short account of what estimation is (p. 215). In the present study, it was up to the children to decide whether or not to estimate, or to infer that it would be necessary to do so.

Responses and contingent questions. For all three tasks, each child was asked to say how many sweets there were (respectively on the plate, in the picture, in the second tube). Two kinds of response were categorized as 'Marked':

- those responses which conveyed vagueness through specific linguistic hedges – 'I think there are ten', 'About ten', and so on;
- vague statements of possibilities or conjectures, conveyed with modal auxiliaries, e.g. 'It might be ten'.

The label 'Marked' and derivative forms will be highlighted in this chapter with a capital letter as a reminder of the interim, technical meaning with which it has been endowed, i.e. referring to the two itemized response types. Hedges and modals will be described jointly as Markers. Children using Markers will be described as Marking, and so on. As already noted, no distinction was drawn in the data-collection phase between epistemic and root Markers. The pragmatic purposes of particular occurrences were analysed, however, at a later stage.

If one of these two kinds of Marker was spontaneously present in the initial response of the child, the interviewer noted it and moved on to the next task (or concluded the interview). Such a spontaneous hedge or modal was denoted a 'primary' Marker. If, on the other hand, the primary response was un-Marked (e.g. 'There are nineteen' or simply 'nineteen'), the interviewer would ask a supplementary question, 'Do you think there are exactly nineteen [or n]?'. This was partly intended to probe the child's commitment to its un-Marked answer; at the same time, my intent was to see whether the children who did not use Marked language spontaneously could be encouraged to do so, thereby revealing that they knew *how* to do so. If this second question provoked a Marker in the child's reply, then this secondary Marker was recorded. Thus, for each of the three tasks, primary and secondary Markers were mutually exclusive.

Data

The interviewer completed a response pro-forma for each child during the course of the interview. Every interview was audiotaped and some were videotaped, so that it was possible to return to the tapes, if necessary, to check the pro-formas and to study prosodic and other nuances of the children's responses. Most of the data relevant to this chapter are shown in Table 3, which gives the number of occurrences of

Table 3   Markers across seven school years in four bands

Year	No.	Primary Markers		Secondary Markers		All Markers	
		No.	% of year	No.	% of year	No.	% of year
Task 1							
R	45	4	9	1	2	5	11
1 and 2	70	1	1	4	6	5	7
3 and 4	65	1	2	7	11	8	12
5 and 6	50	3	6	10	20	13	26
Task 2							
R	45	4	9	1	2	5	11
1 and 2	70	1	1	4	6	5	7
3 and 4	65	6	9	12	18	18	28
5 and 6	50	14	28	10	20	24	48
Task 3							
R	45	1	2	3	7	4	9
1 and 2	70	4	6	4	6	8	11
3 and 4	65	17	26	10	15	27	42
5 and 6	50	11	22	18	36	29	58
All tasks							
R	45	7	16	3	7	9	20
1 and 2	70	5	7	11	16	15	21
3 and 4	65	20	31	24	37	33	51
5 and 6	50	19	38	25	50	37	74

Figure 3   Percentage of children in each age band giving Marked responses to each of three tasks: (a) Task 1, (b) Task 2, (c) Task 3, (d) Tasks 1, 2 or 3

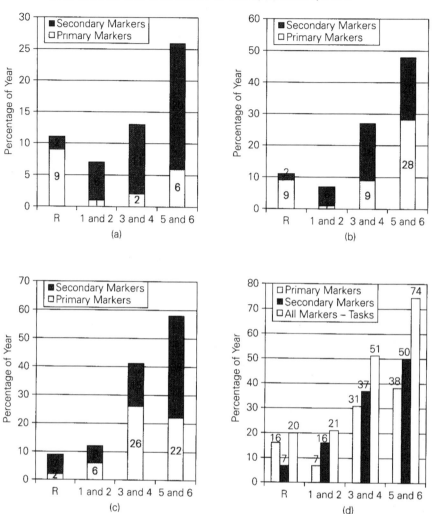

all Markers, separated into the four age bands specified below. These same data are displayed in the bar charts in Figure 3. I shall describe later some additional data (not shown here) which separates the same Markers into finer sub-categories, corresponding to three ability groupings for each age band.

Before proceeding to identify some trends apparent from the graphs, some commentary on the rationale for the organization of the data may be helpful. Compulsory education in England and Wales is organized in chronological 'years', normally beginning (in the absence of nursery classes) at age 4 or 5 with between one and three terms in Year R (for 'reception'). The youngest children in Year 1 will be just 5 at the beginning of the academic year, the oldest nearly 7 at the end. The 'primary' phase of schooling covers Years R to 6.

In the primary school which participated in the study, all children have three terms in Year R. The number of children in each school 'year' (and present for the interview) varied from twenty-three (Year 5) to forty-five (Year R). The results on Marking are presented here in four year-bands rather than seven individual years. The bands are:

- Year R. The first full year in the school, the oldest child being at most 5 years 9 months at the time of the interviews.
- Years 1 and 2. 'Infants', aged between 5 years 9 months and 7 years 9 months.
- Years 3 and 4. 'Lower Juniors', aged between 7 years 9 months and 9 years 9 months.
- Years 5 and 6. 'Upper Juniors', aged between 9 years 9 months and 11 years 9 months.

Interpretive discussion of the data will be located against a background of institutional expectations and indices of arithmetical success within these four bands, these phases of primary schooling. One felicitous consequence of the bandings is to achieve statistically viable group sizes, and a degree of numerical parity between them (in fact, the sub-population sizes are forty-five, seventy, sixty-five, fifty). The numbers of children giving a Marked response to each question are presented in the graphs (Figure 3) as percentages of the number in each band, so that comparisons between the bands may be made.

As I remarked earlier, primary (spontaneous) and secondary (provoked) Markers are mutually exclusive for each of the Tasks 1 to 3 (Figure 3(a)–(c)). Therefore, the respective (dark and pale) columns may be validly stacked, the sum being the total number who Mark their response to that question. Figure 3(d), however, gives the size of the union of the sets (for primary and secondary Markers) in the first three graphs, and so the primary and secondary categories (unions) are no longer exclusive. It would be possible, for example, for a particular child to be recorded as primary on question 1 and secondary on question 3. Therefore these columns are not stacked, but are displayed as three non-additive bars.

Occasional reference will be made to some additional data, collected but not presented in Table 3. This was related to the attainment of the children in the domain of whole number concepts, categorized as below average/average/above average for their school year. This category assignment was entrusted to their class teachers, in recognition of their deep knowledge of the children whom they had taught daily for a whole school year. I regard this as more subjective but more reliable than a spuriously objective measure of attainment obtained, for example, purely from a Standard Attainment Task level. I gave no guidance about the proportion of children to be placed in each category. The relatively greater willingness of the teachers of the oldest (10- and 11-year-old) children to use the extreme categories presumably reflects their perception of their pupils' attainment as realized (or not) rather than potential. For understandable converse reasons, perhaps, the Year R teachers proved reluctant to place their 4- and 5-year-olds in the lower-attaining category.

The Marker data in Table 3 were thus subdivided into three ability groupings for each age band – lower (LA), middle (MA) and upper (UA). Samples at the ability extremes were relatively small in each age band, but any notable differences in Marked language between ability bands will be noted in the commentary which follows.

Observations

My observations here are restricted to some trends and features evident in the graphs. In particular:

- The cumulative (stacked) Marked responses on Tasks 1 and 2 show a drop from the first band (Year R) to the second (Years 1–2) with consistent increases thereafter.
- This cumulative decrease over the first two bands (Tasks 1 and 2) is the result of very clear decreases in primary Marking between those bands.
- With regard to secondary Marking only (Tasks 1 and 2), there is a consistent rise from band to band over the whole age range.
- Likewise, on Task 3, the trend (minor inconsistencies apart) is of consistent increase with age.
- Figure 3(d) indicates a greater tendency towards secondary Markers than to primary ones in the last three bands (Year 1 to Year 6), but for the reverse in the youngest (Year R).
- Taking Figure 3(a)–(c) together (but noting that the vertical scales are different): for the children in the last three bands (Year 1 to Year 6) there is an increasing tendency to Mark the response to each task in turn, i.e. more Mark their response to Task 3 than that to Task 2, and responses to Task 2 are more Marked than those to Task 1. Again, this trend is not evident in the youngest group (Year R).

A note on secondary marking. I have already observed that, for any one of the three tasks, primary and secondary Marking are mutually exclusive, and that this justifies the accumulation of corresponding frequencies in the data and 'stacking' of the corresponding columns in the bar charts. In reading the data, however, it should be borne in mind that secondary Marking is *conditional* on the absence of primary marking. Consider, for example, the results on Task 2: in Years 3 and 4 the 18 per cent secondary Markers are drawn from the 91 per cent non-primaries, and so represent **20 per cent** of those who *could have* given secondary Markers. Whereas in Years 4 and 5 the 20 per cent secondary Markers are drawn from the 72 per cent non-primaries, and so represent **28 per cent** of those who *could have* given secondary Markers. In this sense, the increase (adjusted to 40 per cent) in secondary Marking between the third and fourth bands is even greater than the 'raw' percentages suggest (10 per cent unadjusted). The same conclusion applies to the increase in secondary Markers from the second band to the third on Task 2, and indeed to Task 1 across the last three bands. This 'conditional adjustment' has the *opposite* effect between the first (Year R) and second (Years 1, 2) age bands with regard to Tasks 1 and 2, because there is a sharp decrease in primary Marking between those bands, leaving a larger sub-population available for secondary Marking. For both tasks, however, the increase in secondary marking is equally dramatic, and the inference of a consistent upward trend (the third conclusion above) remains valid.

Interpretive Framework

It is reasonable to suggest that, in a broad sense, the data obtained from the tasks and interviews support the expectation (stated as an *a priori* conjecture earlier in this chapter) that the ability to use linguistic Markers, and the tendency to do so, develops with age – at least over the years of primary schooling. It would not be a gross oversimplification, then, to say that the upward trends in the graphs point to the conclusion that children learn (or acquire increasing facility) to use vague language in order to convey attitudes and points of view, in particular uncertainty, in the primary years.

I shall proceed to propose a sociolinguistic developmental interpretative framework which would account for the upward trend, and which might also accommodate the unanticipated initial drops (from band R to band 1–2) noted above in the first two observations.

I suggest that the modal and hedging linguistic behaviours of the children in each band are related to three fundamental developmental dimensions.

The School as Institution

This first dimension concerns the child's developing apprehension of school – of the roles of the players (particularly teachers and their pupils) in the school situation, and the way they relate to each other in learning situations. By 'apprehension' I mean the totality of factors such as: the child's perception of his or her role and the power vested in teachers; how s/he construes (makes sense of) institutional hierarchies and adult purposes. In a study of English primary school children across Years 1 to 6, Buchanan-Barrow and Barrett (1996) found evidence of developmental trends in the children's understanding of the system of the school, and that the perception of their teacher as a powerful, authority figure emerged over time.

For the child, the social environment of the school may have much in common with the home (e.g. being protected and cared for), but evidently differs from the home in many respects (e.g. the number of individuals in a confined space, the relentless succession of tasks offered). Learning these and other essential differences between home and school is part of what Berger and Luckmann (1967) call 'secondary socialization'. Primary socialization takes place in the home, before school, when the family *is* the world, and reality is circumscribed by the child's experiences within the family. This reality is internalized and contributes to the child's sense of self in relation to a very small number of intimately close significant others.

Secondary socialization involves 'the internalisation of institution-based sub-worlds [. . .] the acquisition of role-specific knowledge' (*ibid.*, p. 158). On going to school[5] the child must learn how to become a pupil, one of many in a class, as distinct from 'a certain mother's child'. The internalization which is part of secondary socialization is weaker than that which takes place in primary socialization; the child cultivates (some better than others) the art of 'role distance', such that s/he is not wholly caught up in the identity which the institution imposes on him or her, but is able to distance him or herself from it (and indeed, to question it) and perceive his or her role and those of others in a detached, formal way. In particular, the child:

> apprehends his school teacher as an institutional functionary in a way he never did his parents, and he understands the teacher's role as representing institutionally specific meanings [. . .] Hence the social interaction between teachers and learners can be formalised. The teachers need not be significant others in any sense of the word. They are institutional functionaries with the formal assignment of transmitting knowledge [. . .] the consequence is to bestow on the contents of what is learned in secondary socialisation much less subjective inevitability than the contents of the primary socialisation process. (*ibid.*, p. 161)[6]

These factors contribute to the child's view of how s/he (and others) are 'positioned' within the institutional practices which characterize schooling. In comparison with the integrated and deeply embedded primary world of the home, in which the child is totally immersed, the world of school is more distant, more utilitarian, *more manipulable* (Woods, 1980, p. 13).

Competence

The second dimension is the child's developing (confidence in his or her) ability to produce desired behaviours within a variety of school practices. Such behaviours include an appropriate response to one's name at registration; reading aloud a limited passage of a book to an adult, with accuracy and minimum delay; giving 'correct' answers to questions (particularly arithmetic ones) to which the enquirer already knows the answer (Ainley, 1988, pp. 93–94; Walkerdine, 1988, pp. 54–55, 89–92).[7] In particular, in the narrow context of the tasks presented to the children in this study, the child progresses over the first few years of schooling from a position where counting is a significant challenge, as far as accuracy is concerned, to one where it is a routine if necessary chore. I shall return to this matter later, with appropriate evidence.

Modality

The third dimension is the child's developing awareness of modal concepts and command of modal language. Hypothetical reasoning is, in the Piagetian formulation of cognitive development, a distinguishing hallmark of formal operational thinking. Such an account would essentially rule out modal concepts for most children in the primary years.

In their account of the growth of possibility notions, Inhelder and Piaget conclude:

> Compared to pre-operational or intuitive thought, concrete operational thought is characterised by an extension of the actual in the direction of the potential. [. . .] [However,] the role of possibility is reduced to a simple potential prolongation of the actions or operations applied to the given context [. . .] They do not consist of imagining what the real situation would be if this or that hypothetical condition were fulfilled, as they do in the case of the adolescent. (1958, pp. 248–251)

Piérrault-Le Bonniec (1980) states the Piagetian position in a way which directly relates to the second and third tasks in the study, suggesting (p. 76) that only at 7 or 8 years do children begin to have some idea of undecidability, and that the ability to reason from hypotheses is not acquired until age 11 or 12. This would be likely to depend, however, on the extent to which the child's thinking was embedded in some real or familiar context (Donaldson, 1978, p. 76).

Moreover, Piérrault-Le Bonniec (1980) has identified the presence and development of 'pragmatic modality' in young (age $3^{1}/_{2}$ to 6) children – evidenced by the ability to assess what could, or could not, be made with a given set of materials. The combined picture, then, is of the development of modality from actions to perceived realities, and thence to imagined worlds. This corresponds to and is consistent with a finding of the Bristol Language Study, that modalized utterances in early English child language predominantly express deontic meanings (actions related to obligation and permission); for example, 'can', 'could' and 'may' are used for action-oriented possibility, 'can' being significantly more prevalent than 'may' (Wells, 1979; Stephany, 1986, p. 390). The first epistemically modalized statements in pre-school children tend to occur about six months later than deontic meanings (*ibid.*, p. 396), but are still extremely rare in comparison.

Some insight into the development of Marked language in the early years of schooling is possible by sifting the results of a large-sample study of infant (age 5 to 7) vocabulary (Edwards and Gibbon, 1973) for the following data.

At age 5+ 'think' is present in the vocabulary, with a lowish frequency index (FI) of 0.3. 'May' and 'might' first appear (in the Edwards and Gibbon sample) at 6+, with FI 0.46 and 1.53 respectively, and 'think' at 0.64. 'Perhaps' does not appear until age 7+, with a low FI of 0.18; by this age the FI's of 'may' and 'might' have roughly doubled in comparison with those at 6+, and that of 'think' has risen to 3.7. 'Maybe' does not appear in the data.

Examination of the Fawcett corpus (Fawcett and Perkins, 1980) produces an interesting, surprisingly consistent account of age-related differences in the use of

Table 4   Age-related differences in the use of Markers

Marker	Age 6	Age 8	Age 10	Age 12
think	8	53	103	130
might	14	27	35	40
may	0	2	5	5
maybe	1	1	4	7
perhaps	0	3	3	5

these Markers. The corpora at each age consists of 15,000 words. Frequencies are as shown in Table 4. Incidentally, data from a corpus of 250,000 English words spoken by American adults (Howes, 1966) contains sixty-six occurrences of 'may', 102 of 'might', 139 of 'maybe' and 1,034 of 'think'.

Of course, 'think' may be used in the root sense, with reference to cognition (as in the imperative 'Think before you speak'). In the Fawcett data, however, the children use 'think' almost exclusively in the epistemic sense, as a commitment marker. For example:

They only have PE once a week, I think. [Age 8]
They're probably for a car or something, I should think. [Age 10]

Against this three-dimensional background I propose the following developmental narrative; the objective is to account for the trends in the data by reference to the interpretive framework that I have sketched.

Interpretation of the Data

The account of changes in Marked language and performance in the primary years (age 4 to 11) is presented below in terms of three developmental phases, which I have termed Initiation, Suspicion, Approximation and Protection. These phases will be briefly characterized, then illustrated by extracts from the 230 task-based interviews.

Year R – Initiation

In this first phase, the child's apprehension of school is relatively naive, and adult behaviour – questioning in particular – is taken at face value, without suspicion. Counting is a relative novelty, and the child is aware that its performance is some-times faulty; this is hardly surprising, given the complexity of the process, as analysed by Gelman and Gallistel (1978), Steffe et al. (1983) and Fuson (1988). Whilst the task of enumerating nineteen items may be *accessible* to a Year R pupil, a teacher will alert the child even when the count is substantially competent yet not entirely accurate; because accurate counting is a major goal, a targeted skill, in this phase of schooling. Moreover, Gelman (1977) suggests that only about one 5-year-old in six can *accurately* enumerate a set of nineteen items, given one minute to do so. In this sample of Year R children, it was actually one in five. The child (Year R) may well wish, therefore, to acknowledge to the interviewer (as a primary Marker) some doubt about the answer s/he gives. In almost every case this is achieved with the plausibility Shield 'I think', with just a few epistemic modals, 'may', 'might' and 'maybe'. The child's Marking can be understood as straightforward co-operation, in effect observing that the maxim of Quality is met, or serving notice that it may not have been met (Brown and Levinson, 1987, p. 164).

Tasks 2 and 3 were invariably approached by attempting a count of the sweets that are visible in the picture/tube. Proportional reasoning (Task 3) and attempts to compensate for the possibility of hidden sweets (Task 2) were rare. The request by the interviewer for the number of sweets was accepted as achievable rather than as a 'trick' question about something indeterminate or hypothetical.

It was the case that two-fifths of the Year R children who had (in advance) been informally assessed by their teachers as 'above average for the year' with regard to whole number concepts were recorded as giving a Marked response, whereas only two 'average' and no 'below average' Year R children used any Markers. This is not easy to interpret. It could mean that the lower attainers prefer not to draw attention to their uncertainty, or lack the linguistic means to communicate it. Perhaps they want the confidence to discuss their answer with the interviewer, or even lack awareness of its unreliability.

In each of the following transcripts 'I' is the interviewer, 'C' the child.

Example 1   Boy aged 5:9 (M39)

The young boy in this transcript uses Marked language fluently.

M39:1	*I:*	Can you tell me how many sweets there are on the plate?
		[C counts aloud, points to each sweet in turn, takes 17 seconds to reply]
2	*C:*	One, two, three, four, five, six, seven, eight, nine, ten, eleven, twelve, thirteen, fourteen, fifteen, sixteen, seventeen, eighteen, nineteen, twenty, twenty-one, twenty-two . . . twenty-three . . . twenty-four.
3	*I:*	Twenty-four? Do you think there are exactly twenty-four?
4	*C:*	**Maybe** not.
5	*I:*	Why do you say that?
6	*C:*	Because I **might** have counted two double.
7	*I:*	[indicating photo] Right, now, there are some sweets in the glass. Can you tell me how many sweets there are in the glass?
		[C counts aloud, touches the sweets in the photo, takes 6 seconds to reply]
8	*C:*	One, two, three, four, five, six, seven, eight, nine.
9	*I:*	And do you think there's exactly nine?
10	*C:*	**I don't know. Maybe** there's one more, because **I don't know** if I counted that one.

11 *I:* OK. Now, before we went into assembly I put ten sweets in that tube, so I know there are ten sweets in there. Can you tell me how many sweets there are in that tube?
[C counts aloud, touches the outside of the tube, no comparison of the tubes, takes 22 secouds to reply]

12 *C:* One, two, three, four, five, six, seven, eight, nine, ten, eleven, twelve, thirteen, fourteen, fifteen, sixteen, seventeen, eighteen, nineteen . . . twenty, twenty-one, twenty-two, twenty-three, twenty-four, twenty-five.

13 *I:* Twenty-five. And do you think there are exactly twenty-five?

14 *C:* **Maybe** not, because there's so much.

The 'How many?' question is received by this boy as a straightforward invitation to count. He has command of the stable order principle (M39:2), but applying the one–one principle to a disordered set of this size presents difficulties for him. He does not touch the sweets one by one, but points to them – a partial internalization. His ability to partition the set (Gelman and Gallistel, 1978, p. 77) into counted and to-be-counted subsets is faulty, *and he knows* it is sometimes faulty (M39:6). In fact, he has double-counted five of the sweets. His awareness of the same difficulty, with partitioning, prompts his Marked response (M39:10) to Task 2. In this case he is suspicious of omission, whereas before it was duplication. For Task 3 his order stability in reciting the 'tags' 1 to 25 is once again faultless; he expresses his uncertainty in his answer with the same epistemic hedge (M39:14) 'maybe' (not). The meaning behind his voluntary explanation 'because there's so much' is unclear, but could suggest a tacit estimate of some sort which does not accord with his count. Alternatively, he may in effect be acknowledging that he regularly makes a partitioning error when the set to be counted is as large ('so much') as this one.

In each case, he applies the cardinality principle unerringly. In fact, his competence on the first and third tasks confirms Fuson's ordering of mastery of the principles for sets of this size. What is very clear from this transcript is that he takes a face-value view of the interviewer's three questions, that he counts rather than estimates, but that he is quite uninhibited about notifying the interviewer that his (the child's) answers may be unreliable. His attitude is something like: 'That's what I make it but I know from experience that I may have made an error. That's simply the way things are when, like me, you're a novice at counting.'

Years 1 and 2 – Suspicion

The child's apprehension of school includes the sense of being scrutinized by curious adults, of the existence of 'testing' questions. S/he is now expected to enumerate small sets routinely, and has built on this for the purposes of addition and subtraction of whole numbers. The result is a manifest reluctance to use primary Markers in response to the first two tasks (perhaps because 'he wants to know if I can get it right'), but the interviewer's probe ('Exactly?') may release an acknowledgement of uncertainty, using the same Plausibility Shields (predominantly 'I think', with a few 'maybe's) as the Year R child. The more subtle linguistic ability to use Approximators as epistemic hedges, for self-protection by vagueness, has not yet been developed.

Task 3 may seem bizarre, quite unlike routine practical or textbook counting exercises; it gives rise to more hedged responses, though not in the low attainers, who are most likely to hedge on Task 1 – like the younger, higher-attaining Year R children.

Example 2   Boy aged 7:5 (M55)

The transcript is chosen for absence of Marked language.

M55:1   *I:*   Can you tell me how many sweets there are on the plate?
            [C counts aloud, touches each sweet, takes 23 seconds to reply]
     2   *C:*   [quickly] One, two, three, four, five, six, seven, eight, nine, ten, eleven, twelve, thirteen ... oh. [Restarts counting, now more slowly, placing sweets on the table whilst counting] One, two, three, four, five, six, seven, eight, nine, ten, eleven, twelve, thirteen, fourteen, fifteen, sixteen, seventeen, eighteen, nineteen.
     3   *I:*   And do you think there are exactly nineteen?
     4   *C:*   What?
     5   *I:*   Do you think there are exactly nineteen?
     6   *C:*   Er ... [pause, then recounts into hand] One, two, three, four, five ... One, two, three, four, five, six, seven, eight, nine, ten, eleven, twelve, thirteen, fourteen, fifteen, sixteen, seventeen, eighteen, nineteen.
     7   *I:*   So do you think there are exactly nineteen?
     8   *C:*   Yes.
     9   *I:*   Right ... can you tell me how many sweets there are in the glass?
            [C turns the photo over to try and see the 'back' of the glass; counts aloud, takes 20 seconds to reply]
    10   *C:*   Oh! One, two, three, four, five, six, seven, eight, nine, ten, eleven. ... one, two, three, four, five, six, seven, eight, nine, ten, eleven, eleven.
    11   *I:*   And do you think there are exactly eleven?
    12   *C:*   Yes.

His first recount (M55:2) is presumably a self-correction when he suspects that his partitioning has gone wrong. There is no problem with the order of the tags, and the count is in fact accurate. The interviewer's probe (M55:3, 6) is immediately taken to be a suggestion that he has miscounted. Instead of hedging, he recounts the set (M55:6). The indeterminacy of the number of sweets in the glass is either not perceived or not acknowledged (M55:12). Again, he counts the (visible) sweets unprompted.

This child's uncertainty, his sense that he may make mistakes when he counts, is conveyed not in Marked language but by his inclination to recount. His approach to Task 3 (below) is much the same, until he admits some (unspecified) difficulty. 'Can't' (M55:14) is, of course, a root modal associated with capability.

The interviewer's prompt (M55:15) precedes proportional thinking and an un-Marked response.

13 *I:*   Now, I counted some sweets into here, and I know there are ten sweets in there [...] Can you tell me how many sweets there are in that tube?

14 *C:*   One ... two ... three ... four ... Can't do this! One, two, three, four, five, six, seven, eight ... One, two, three, four, five, six, seven, eight, nine, ten ... nine, ten ... eleven ... one, two ... I can't figure it out right.

15 *I:*   How many do you think there are? [Pause] There are ten in there, OK [...] We know that there are ten in there.

16 *C:*   I know that ... [compares tubes without counting] ... twenty.

17 *I:*   Why do you think that?

18 *C:*   Because it's half.

19 *I:*   OK. Excellent [...]

Years 3 to 6 – Approximation and Protection

There is developmental continuity within this phase rather than qualitative change. The account which will follow characterizes a child who has moved some way along that developmental continuum. In fact, the 'low ability' children in this phase used Markers consistently less (significantly less on Task 3) than their higher-attaining peers – in many respects their response was more like that of the younger, higher-attaining Years 1–2 children.

The child has a well developed apprehension of his or her role in the practice of education. S/he realizes that teachers' – including researchers' – questions about mathematics are usually not simple requests for information (Walkerdine, 1988, p. 61). Indeed, s/he may have become quite expert at eliciting from the teacher the very information that s/he was originally asked to produce, or an easier question with the same answer (MacLure and French, 1980). This kind of pupil behaviour is a parody of 'thinking', a strategy with a long pedigree:

> Each time I had to think of a question easier and more pointed than the last, until I found one so easy that she would feel safe in answering it [...] In fact, she was not even thinking about it. She was coolly appraising me, weighing my patience, waiting for that next, sure-to-be-easier question. I thought, 'I've been had!' The girl had learned how to make all her previous teachers do the same thing. (Holt, 1969, p. 38)

On the other hand, the child is confident in its ability to count; of the 108 Year 3 to 6 children who did choose to count the nineteen sweets in Task 1, about two-thirds did so accurately, and four-fifths of the remainder obtained eighteen or twenty. If such a child does in fact count the sweets in Task 1, a primary Marker is unlikely. On the other hand, the child may judge that the interviewer is not interested in the precise number of sweets on the plate, and offer a primary Marked estimate (there is a corresponding rise in the fourth band), or a response which is tagged by intonation, with a question mark.

The child recognizes that some quantities are indeterminate; on Tasks 2 and 3 s/he will realize, despite the awareness of 'testing', that the interviewer cannot sensibly expect precise answers to these 'How many?' questions. If the child's first answer is un-Marked – as if s/he were guessing how many sweets in a jar at a fête – it will readily admit to uncertainty when asked if that answer is exact. The child may recognize the proportional reasoning Task 3 for what it is, and (correctly) have some confidence that there are exactly twenty sweets in the second tube. This is strongly reminiscent of many 11-year-olds involved in Assessment of Performance Unit practical testing, who were very resistant to the notion that a 20 g weight was being counterbalanced by twenty-one (rather than twenty) plastic tiles (Joffe, 1985, p. 21A). These pupils tended to 'demand a recount' in order to obtain the preferred 'round' answer. At the same time, however, Clayton observes that pupils will frequently actually avoid round numbers when asked to estimate numerosity. He calls this tendency the 'jelly-baby effect':

> Guess the number of jelly-babies in the jar and whoever is closest wins the prize. Some pupils appear to believe that the person in charge would not have a 'round' number of jelly-babies, so they guess a number close to but not exactly the round number. (Clayton, 1992, p. 117)

At the same time, the child has developed the competence to deploy Approximators such as 'about' (as well as modal auxiliaries) as epistemic markers, to introduce vagueness for protective purposes.

Example 3   Boy aged 8:1 (M165)

M165:1	*I:*	Can you tell me how many sweets there are on the plate?
		[C points to each sweet, counts silently, takes 19 seconds to reply]
2	*C:*	Nineteen.
3	*I:*	Do you think there are exactly nineteen?
4	*C:*	Yes.
5	*I:*	And how do you know that?
6	*C:*	I counted up in twos.
11	*I:*	OK. Can you tell me how many sweets there are in the glass?
		[C points to sweets, counts silently, takes 20 seconds to reply]
12	*C:*	Ten.
13	*I:*	Ten. And do you think there are exactly ten?
14	*C:*	No.
15	*I:*	No? Why not?
16	*C:*	Mm . . .'Cos there **might** be some more at the other side.
17	*I:*	I've put ten sweets in this tube here, OK? . . . Can you tell me how many sweets there are in that tube?
		[C touches (turns tube), counts silently, compares tubes, takes 31 seconds to reply]
18	*C:*	**About** twenty-one.

Quite confident of his internalized count of the sweets, he has no cause to Mark his responses (M165:2, 4, 6] to Task 1. On the other hand, the indeterminacy of Task 2 gives rise to the secondary epistemic modal 'might' (M165:16). His approach to Task 3 is cautious and unhurried. The interviewer judged that some comparison was drawn between the tubes and hence that there may have been some proportional inference. The eventual answer (M165:18) includes an Approximator, 'about'. It is not possible to judge from the transcript whether this hedge is epistemic or root. The actual estimate of twenty-one (which is not a 'round' number) appears to be an instance of the 'jelly-baby effect' influencing the choice of estimate.

Example 4   Girl aged 10:6 (M222)

M222: 1	*I:*	Can you tell me how many sweets there are on the plate?	
		[C doesn't count, takes 2 seconds to reply]	
2	*C:*	**About** twenty?	
3	*I:*	Now, can you tell me how many sweets there are in the glass?	
		[C doesn't count, takes 1 second to reply]	
4	*C:*	Ten.	
5	*I:*	And do you think there are exactly ten?	
6	*C:*	No! [laughs] **Not exactly**.	
7	*I:*	Why did you say that?	
8	*C:*	'Cos it's **not actually** . . . it doesn't look like ten . . . Well, I just guessed.	
9	*I:*	OK. Now, I've put ten sweets in this tube . . . Can you tell me how many sweets there are in this tube?	
		[Stares, compares tubes, takes 1 second to reply]	
10	*C:*	Twenty.	
11	*I:*	And do you think there are exactly twenty?	
12	*C:*	**About** twenty-five . . . **or** twenty.	
13	*I:*	And what makes you say that?	
14	*C:*	'Cos it **looks like** half, twice as much as in there.	

This pupil approaches each task with mathematical confidence and social maturity. On the video she looks relaxed, at ease with the interviewer. She knows exactly when an estimate will suffice and, indeed, when nothing else is possible. For Task 1, she instantly judges that an estimate will meet the requirement of the maxim of Quantity. Her Approximator qualifies a suitably round number (M222:2) with the force of a root hedge (as if to say, 'This is as much as you need to know'). She just laughs at the suggestion that her vague use of ten (M222:4) as a 'cognitive reference point' (Rosch, 1975) should be taken to be anything other than an approximation. For Task 3, she is explicit that proportional reasoning is the basis of her rounded estimate.

Her response (M222:12) to the interviewer's probe (M222:11) needs closer examination. Channell's analysis of the approximator form *n or m* (1994, pp. 53–58), drawing on Crystal (1969), reveals a subtle prosodic distinction between a binary alternative and a vague range of alternatives. Consider the difference between the two following questions (scenario: parent to child about to go on a school trip):

I:7.1 Will you need five or ten pounds?
       [Meaning: which of two banknotes shall I give you?]

I:7.2 Will you need five or ten pounds?
       [Meaning: is that the kind of amount you'll need?]

The semantic distinction is achieved by prosodic marking of the questions. In the first case, the two alternatives (*five, ten*) are both stressed and nuclear (i.e. have maximal prominence). In the second, *five or ten* is a single, unstressed tone unit and *pounds* is nuclear. Channell also points out that, for the approximative use of *n or m, n* is always less than *m*. This is not the case in (M222:12). Sure enough, replaying the videotape of the interview confirms that the alternatives *twenty-five, twenty*, are not parts of a single tone unit in the utterance. They are quite separate, the first being stressed slightly more than the second. The impression conveyed is that she considers revising her estimate (to twenty-five, the next 'round' number in this range) but changes her mind perhaps on grounds of comparison and proportional reasoning.

Gender Differences

An expectation that girls/women will hedge more than boys/men is common (Lakoff, 1975). The expectation is related to a supposition that males are more assertive and confident than females in a number of social situations. Clayton (1992) identifies estimation as a risk-taking school activity, and concludes that boys, on the whole, approach it with greater confidence than girls, particularly in the years of secondary schooling. In so far as modals and hedges can achieve a measure of protection for the speaker against accusation of error, one might expect some tendency for the girls in the whole-school study to use these Markers more than boys in their responses to the three tasks.

The data from this study do not, in general, bear out the expectation. It turns out that one-fifth of infant (4–7) boys used some Marked language, as did one-fifth of infant girls. For junior (7–11) children, the proportion was three-fifths for both boys and girls, with 41 per cent of boys Marking one or more responses ($n = 125$) and 41 per cent of girls ($n = 105$) over the whole school population.

Since a global report on gender differences (or the lack of them) may overlook some fine distinctions, I looked for differences of a more detailed nature. Expecting to find such differences, if any exist, among the junior pupils, I concentrated my efforts on the 115 pupils in Years 3 to 6. Once again, the proportions of boys and girls exhibiting the linguistic behaviour were much the same in every case. For example see Table 5. Clayton's findings would lead one to expect that gender differences in Marked language might become more evident in the years of secondary schooling (after age 11).

Table 5   *Boys' and girls' responses (per cent)*

Linguistic behaviour	Boys	Girls
Giving at least one primary Marked response	36	31
Giving at least one secondary Marked response	42	43
Giving at least one Marked response to Tasks 2 or 3	56	57

Modal Auxiliaries

The term 'Marked' has been introduced to include modal auxiliaries and hedges. The two classes of language clearly overlap in adverbial forms such as 'possibly', 'maybe'. Only nineteen children in the sample of 230 used modal language that was not also recorded as a hedge. These nineteen were predominantly in the third age band (Year 3 and 4) and in nearly every case the modal verb used was 'might'. The sole exception was a girl in Year R, aged $5\frac{1}{2}$, judged to be mathematically above average, who made a fully internalized count of the sweets in response to each task (with no external manifestation of the two 'local correspondences'). Her eventual answer to each of the three 'How many?' questions was itself marked as a question, with rising intonation, but with no secondary hedge on the first two tasks. For the third she gave the answer twenty-three, and in reply to the prompt 'Do you think there are exactly twenty-three?' she answered, 'There may be one or two more,' with prosodic marking indicating epistemic approximation.

Of the remaining eighteen, all of whom used 'might', the following are typical:

I might have counted two double. (Year R boy, Task 1)

There might be some more. (Year 1 girl, Task 2)

There might be some more on the other side. (Year 3 boy, Task 2)

There might be less. (Year 3 boy, Task 3)

It might not be doubled. (Year 4 girl, Task 3)

It is interesting to note that, whereas occurrences of 'may' and 'might' were rare, in every case they arose as *secondary* Markers, in response to the interviewer's prompt: the examples above clearly convey this reactive uncertainty.

The adverbial hedge/modal 'maybe' is also used by thirteen children (three of whom also use 'might'), and is more evenly distributed over the age range in the school.

Prosody

A decision (arguably a conservative one) was made not to record as a hedge a third type of response, namely uncertainty expressed in assertions which are offered as questions, usually through rising intonation (such as 'Twenty?'). I am acutely conscious that such responses are contenders for inclusion, as Shields. Stubbs (1986, p. 21) observes that tag questions 'allow statements to be presented as obvious, dubious, or open to challenge'. What is true in this respect of explicit tags (such as '. . . , isn't it?') must also be true of tacit, prosodic tag questions.

Of the 230 children in the sample, fifty-two gave numerical answers marked by rising intonation; of these thirty Marked one or more of their responses in some other way which was included in the analysis. The remaining twenty-two remain outside the modal/hedge consideration of this study. The oldest year-group (Year 6) accounts for none of the twenty-two, who are fairly evenly distributed across the earlier Years R to 5. Including these in the 'Marked' category does not affect general trends, owing to the conspicuous rise in Marked responses in the fourth band (Years 5 and 6) under the operational definition used in the study.

Summary

In this chapter, I have presented and interpreted data obtained from short, task-oriented interviews with the pupil population of one primary school. I began with the expectation that the data would support a hypothesis that the ability to use linguistic Markers, and the tendency to do so, increases consistently through the primary years. These Markers (hedges and modal auxiliaries) serve epistemic and root purposes, conveying the speaker's uncertainty or awareness that an estimate was appropriate (in fact, essential in the case of the last two tasks). The outcome of the study includes an account of some developmental aspects of conveying uncertainty in mathematics, including social aspects which would account for a dip in unprompted Marked language shortly after the child's initiation into schooling.

I have argued that, in the context of mathematical activity, uncertainty is a normal state, potentially a creative one. A creative objective for teachers could be to recognize it and work with it (as well as with their own epistemological and pedagogical uncertainties) rather than to seek – in the short term, at least – to deny or eliminate it.

Whilst it is clearly the case that the use of Marked language has, in some sense, to be learned, it is not so clear how that particular linguistic competence is acquired, a competence that includes a modal dimension which lies at the core of the communication of propositional attitude. The unexpected outcome of this particular study, in the context of estimation activity, is that children may be socialized into suppressing this aspect of their linguistic competence until they discover, or assert, that – in some mathematics classrooms, at least – it is all right to be wrong.

Notes

1   Van den Brink (1984) argues that the linguistic component, which he calls 'acoustic counting', develops and functions in young children quite independently of any reference to objects. He shows, moreover, that acoustic counting is free of some of the constraints of conventional 'quantity counting'.

2 I am grateful to Heather Cooke for articulating this suggestion, at a research seminar on vague quantitative language, given by Joanna Channell at the Faculty of Mathematics, Open University, Milton Keynes, on 14 June 1993.

3 The use of actual Smarties was vetoed by the interviewer, on the grounds that Smarties are a product of Rowntree's, which is a division of Nestlé, which in turn is the subject of a boycott intended to reverse that company's promotion of baby milk products in the Third World.

4 In the pilot study these and other forms of the questions were trialled. Notably the form 'How many sweets do you think . . . ?' was trialled, despite unease that it might prompt a bias towards 'I think . . .' responses. In the event, there was no evidence from the trial to suggest that it did have that effect. We also considered simply asking 'How many sweets are there . . . ?' but rejected it without trial on the grounds that it came over as too direct, somewhat aggressive and 'testing'. Only later did we rationalize the guidance of our intuition in preference of presenting the question as an indirect speech act, whose illocutionary force is discussed again in Chapter 8 (case 1: Hazel).

5 The social and economic structure of Britain has changed significantly since the publication of Berger and Luckmann's account. There are certainly home environments to be found in the 1990s which do not offer the same secure and cosy base for primary socialization as the stereotype described by Berger and Luckmann. Furthermore, the processes of secondary socialization may begin before formal schooling in other environments such as a childminder's home or a playgroup.

6 Reference to teachers as 'functionaries' is quite shocking. In many countries they are literally that in the sense that they are civil servants. Perhaps the epithet is now largely apposite in England and Wales, since the state has assumed such tight regulation of the government of schools, of the curriculum, and is now embarking on national specification of pedagogy. Teachers, like milkmen, have been cast in the role of 'deliverers' of a product.

7 Walkerdine (1988, pp. 89–92) points out that the form of the pedagogic discourse between mothers and their children, and that between teachers and their pupils, is remarkably similar, and that both frequently use pseudo-questions. She argues, however, that teachers are principally observers, operating within a testing regime, whereas mothers are participants in the activities which are the discourse contexts.

8 Pragmatics, Teaching and Learning

> Language which fulfils a genuine need ... will ultimately find an audience ... to whom it is not obscure. (Wallwork, 1969)

Whereas the data in Chapters 4 to 6 are unified personally by informant (e.g. Susie) or thematically by phenomenon (e.g. hedges), those to be presented and considered here are more diverse in origin and nature. There is another important distinction: whereas the data in the earlier chapters were collected in the course of research interviews, the data presented here are transcripts of a variety of teaching and learning situations. Many of these transcripts were supplied by members of an 'Informal Research Group' of school teachers which met to consider the place of vagueness in mathematics talk, and the implications for their interactions with pupils.

I shall apply the complete linguistic toolkit assembled in the book to the analysis of a number of 'cases' – transcripts or fragments – demonstrating further the role of vagueness in mathematical discourse by reference to a number of episodes in teaching and learning. These cases come from a variety of sources, including the Informal Research Group.

The Informal Research Group

In the autumn of 1994, I broadcast an invitation to primary and secondary school teachers to join an 'Informal Research Group' which I would convene early in 1995 to consider aspects of vague language and mathematics. About a dozen teachers indicated interest. I wrote to them to indicate the intended focus of the group in more detail, concluding:

> Within the informal group I should like to share my findings to date, and in time to encourage group members to collect further examples of the same kind of language in mathematics classroom. Such examples might be generated in individual or small group discussion with you, or in whole-class discussion. [...] The important factor would be that your examples are collected in the context of 'normal' classroom teaching.

One useful aspect of this group was the age-range spread which it represented – from a reception class teacher to another teaching Advanced-level students in a sixth-form college. One member of the group was in her first year of teaching, two were in their third, and the others had been teaching for some years. We rapidly discovered that the language of classroom mathematical discourse provided a focus for our discussion and close attention, a focus that facilitated and energized mutual effort and understanding. Indeed, the shared cross-phase dimension added greatly to the interest of the work of the group.

At our first few meetings I shared my own understanding of the functions of vague language in mathematics talk with reference to some transcripts of my mathematical conversations with children. I distributed papers for further reading, and we discussed possibilities for tasks that would be an appropriate basis for contingent mathematical discussion between them and their pupils. Each member of the group then undertook to carry out, record and transcribe at least one such conversation in their own classroom. After further meetings for preliminary reporting back, their transcripts were ready.

So much for the background and source of these data. Equipped with a set of approaches to discourse, I now proceed to the examination of eight 'cases'. The teachers featured in the first five were members of the Informal Research Group. Some cases are based on quite lengthy transcripts, and I am obliged here to be selective. The purpose of the analysis is to illustrate a variety of approaches to such discourse, and to give evidence of consistency with the observations from my own contingent interviews with children, as presented in the earlier chapters.

Case 1: Hazel

Hazel is a primary school teacher in her third year of teaching. In the transcript (IRG3) she is talking to two 10-year-old girls in her class, Faye and Donna. Hazel indicates that 'both are able mathematicians who often work together'. The conversation is an exploration of the difference between b^2 and ac, where a, b, c are consecutive terms of an arithmetic sequence. It falls into four episodes:

- *Episode 1*. Investigation of the case when the common difference is 1 (IRG3:1–61).
- *Episode 2*. Investigation of the case when the common difference is 2 (IRG3:62–105).
- *Episode 3*. Investigation of the case when the common difference is 3 (IRG3:106–120).
- *Episode 4*. Search for a higher-level expansive generalization (Harel and Tall, 1991: see Chapter 2) which includes the three generalizations arrived at inductively in the previous episodes as special cases (IRG3:121–160).

First, observe that in every case Hazel's instructions and requests to the two girls are presented as indirect speech acts, for example (there are many):

17   *Hazel:* Shall we try it out and see what happens? Do you want to each choose your own set of consecutive numbers?

66   *Hazel:* Right. Would you like to try out with ten, twelve and fourteen, one of you, and the other one can try another jump.

130 *Hazel:* Can you tell me what the difference in the answers of the two sums
that, the two multiplications you're doing, would be when you have
a difference of four between each number?

IRG3:17 and 66 are on-record FTAs, with redressive action ('orders' presented as
questions) with regard for the children's negative face, as Hazel imposes on their
personal autonomy of action. These are conventionally polite, indirect speech acts
(like 'Can you pass the salt, please?'). She believes that the investigation will be a
worthwhile, educative experience for them with a potentially stimulating outcome.
Nonetheless she recognizes the risk-taking which is inherent in her quasi-empirical
approach, and that she requires their co-operation as active participants in the
project as they generate confirming instances of generalizations-to-come. In IRG3:17
she says, 'Shall we try it out?', the plural form including and identifying herself as
a partner in the enterprise. In IRG3:130 she probes for a prediction (related, possibly, to an as yet unarticulated expansive generalization) and realizes the threat to
the girls' positive face – what if they fail to make a correct prediction, will their
reputation as 'good mathematicians' be dented? IRG3:130 respects their positive
face, and the indirect modal form redresses the on-record FTA. The form of her
question – 'Can you tell me . . . ?' – is precisely that which I chose for the whole-
school developmental study reported in Chapter 7. Quasi-empirical teaching, inviting conjectures and the associated intellectual risks, is unimaginable if the teacher
is not aware of the FTAs that are likely to be be woven into her/his questions and
'invitations' to active participation. Redressive action dulls the sharp edge of the
interactive demands that this style places on the learner. For Hazel, notwithstanding
her authority in her own classroom, the indirect speech act has become a pedagogic
habit, which even extends to non-cognitive requests:

41 *Hazel:* Would you like to go and get one [a calculator], Donna?

Children are likely to be less sensitive, to be more direct:

42 *Faye:* Get two.

The children make frequent use of Shields as epistemic hedges. Early in the conversation Faye (IRG3:9) observes a difference of 1 between 10×12 and 11^2. Somewhat precipitately, perhaps, Hazel asks:

10 *Hazel:* One number difference . . . Do you think that will always happen
when we do this . . . ?

Faye readily agrees, but Hazel, perhaps realizing that she has not probed but has
'led the witness', seems to want to give them more of an option to disagree.

12 *Hazel:* What makes you think that? Just 'cos I asked it . . . or . . . ?

Donna gives hedged agreement (14), and Hazel invites her (15) to account for her
provisional belief.

14 *Donna:* I think so.
15 *Hazel:* Why?

Arguably this is a tough question – to account for a belief that one is not really committed to anyway. Donna's justification (16) is phenomenological rather than structural.

16 *Donna:* Well, if, um . . . if it's after each other like ten, eleven, twelve . . . um . . . it will be one more, because it's one more going up.

It is the basis of a subsequent expansive generalization at the beginning of Episode 2.

62 *Hazel:* Ok. Right, what would happen if you had numbers that jumped up in two instead of one, so you had ten, twelve and fourteen?
63 *Faye:* I think the answer is a two-number difference. So two.
64 *Donna:* Yeah, yeah. So do I.

The substantive proposition in IRG3:62 – that there is a two-number difference – is, in fact, false. By prefacing it with an epistemic hedge, Faye marks her utterance as a conjecture and withholds commitment to it.

Returning to Episode 1: Hazel encourages the children to try out two more examples with three consecutive integers. They obtain a difference of 1 in each case and Faye (27) affirms her belief (unhedged) that, as Hazel puts it (26), 'that will always happen'.

26 *Hazel:* Do you think that will always happen, then?
27 *Faye:* Yes.
28 *Hazel:* How can you say for certain, 'cos you've only tried out three examples?

When pressed by Hazel to account for her belief (33), Faye attempts a start ('Because . . .') and then backs off (34):

33 *Hazel:* . . . why do you think that for certain?
34 *Faye:* Because . . . well, I don't know for certain but I think . . . 'cos the numbers that we've done are quite close to the first . . .

Abandoning her attempt to respond to Hazel's request for an explanation, Faye's 'well' (34) is a maxim hedge; she realizes that her reply will not be fully co-operative. She cannot sustain the suggestion that she is certain, and is obliged to refuse Hazel's request. Her reply (a second part pair) is dispreferred in relation to Hazel's first part, and is marked by hesitation (Chapter 6). Indeed, her subsequent 'the numbers that we've done are quite close to the first . . .' is vague, hedged with an adaptor ('quite') and violates the maxims of Quantity and Manner. Her 'well' suggests that Faye had foreseen the inadequacy of her explanation.

Donna offers a brief diversion:

35 *Donna:* I don't think it will happen if you do like eleven, fourteen, twenty-two.
36 *Hazel:* But you're talking about the one that . . . if you always have a set of three consecutive numbers will it work?

Her 'like eleven, fourteen, twenty-two' is a delightful example of a vague *generality* (in the sense of Peirce, 1934: see Chapter 3). It is the interpreter's task to determine what it points to, what is included by it. It is difficult to judge how Hazel interprets it, except that she takes it to exclude 'three consecutive numbers' – and perhaps that is precisely what Donna intended to convey through her example. Evidently 'consecutive' is a useful but neglected item in the mathematical lexicon.

As in my conversations with Susie (Chapter 5), the pronoun 'it' is deictic in IRG3:35–36, anaphoric in fact, and co-referential with an earlier demonstrative pronoun, 'that':

26 *Hazel:* Do you think that will always happen, then?

Again, 'it' (35, 36) (and 'that' (26)) has no conventional name, but Hazel, Donna and Faye tacitly understand its deictic referent – a proposition which Hazel has come closest to articulating, much earlier in the conversation.

10 *Hazel:* One number difference . . . do you think that will always happen when we do this where we've got three consecutive numbers and we multiply the two end ones . . . and then in the middle?

Faye brings the discussion back on course with a request for a crucial experiment (Balacheff, 1988: see Chapter 2).

38 *Faye:* I'd like to try it out in the hundreds.

Donna's choice for the experiment seems to be guided by Hazel:

39 *Hazel:* [to Donna] You want one difference between each of those. If you're going to start with a hundred you could have a hundred and . . ., a hundred and one and a hundred and two. Would you like a calculator . . . ?

Faye's independent choice of 'any old' set of three consecutive integers – 110, 111, 112 – becomes apparent later (60):

51 *Faye:* I still get one number different.
52 *Hazel:* So that . . . so do you . . . will it always work, d'you think?
53 *Faye:* Yeah . . . I think.

54	*Hazel:*	How can you be sure?
55	*Donna:*	Umm . . .
56	*Faye:*	[laughing] Well . . .
57	*Hazel:*	Are you sure?
58	*Faye:*	Well not really, but . . .
59	*Donna:*	Quite, yeah.
60	*Faye:*	I think so. Yeah, quite sure. Because it has worked because we've done ten, eleven . . . Well I've done ten, eleven, twelve, nine, ten, eleven, which are quite similar, and then I've jumped to, um, um . . . a hundred and ten, a hundred and eleven and a hundred and twelve. It's quite a big difference. So yeah?
61	*Donna:*	Yeah, so do I.

By this stage Hazel seems reluctant (52) to influence their commitment to the generalization (the 'it' that 'always works'). Faye's intellectual honesty is very evident here. Her crucial experiment (60) provides another (presumably weighty) confirming instance of the generalization (51), yet her assent to it is still hedged, partial (53). One senses that Hazel has created, or nurtured, a Zone of Conjectural Neutrality in which Faye understands that it is the conjecture ('it always works') which is on trial, not her. She is free to believe or to doubt. Nevertheless, her 'well's (56, 58) indicate dispreferred turns; she senses, perhaps, that it would be easier if she agreed, that agreement would better respect Hazel's positive face wants – for Hazel would gain satisfaction from Faye's coming-to-know. At the end of Episode 1 she goes some way towards agreement (60), affirming in the end that she is 'quite sure', i.e. even more sure than might be expected, the Adaptor 'quite' making things 'less fuzzy' (Lakoff, 1973, p. 471). She proceeds (60) to reflect in detail on the variety of evidence which she has assembled, to account for her willingness to make the enthymematic leap into the unknown. Donna is apparently something of a passenger compared with Faye.

I conclude this selective study of Hazel's transcript by considering Episode 4, which opens as Hazel invites the two girls to 'recap' on their generalizations to date:

121	*Hazel:*	Ok. So if you recap. Would you like to start from the beginning and tell me what the difference is when you've got a jump of one, what the difference is in the answers. When you've got a jump of two what the difference is in the answers.
122	*Faye:*	Ok. When you've got a jump of one the difference in the answers is one.
123	*Donna:*	When you've got a jump of two the answer . . . the difference in the answer is four.
124	*Faye:*	And when you've got a jump of three the difference in the answer is nine.

The children observe that each difference is a square (125–129). Next, Hazel invites a prediction:

130	*Hazel:*	Can you tell me what the difference in the answers of the two sums that, the two multiplications you're doing, would be when you have a difference of four between each number?
131	*Donna:*	Twenty . . . twenty-six.
132	*Hazel:*	You think twenty-six, Donna. What do you think, Faye?

Hazel's response (132) to Donna's prediction (incorrect, as it happens) is of interest to me, since I observed the same linguistic behaviour in myself (Chapter 6) – the use of an Attribution Shield to sustain the involvement of the other child. In this case Donna corrects herself, and Faye agrees –

| 133 | *Donna:* | [interrupts] No, sixteen. |
| 134 | *Faye:* | Sixteen, yeah, sixteen. |

Donna's error (131) was perhaps computational ($4^2 = 26$) rather than algebraic (the difference is 4^2).

Next, Hazel invites predictions for differences of 5 and for 6. Donna and Faye give responses of 25 and 36. Hazel moves to the expansive generalization (146–155):

156	*Hazel:*	What's the pattern, then? Can you sort of explain the pattern for me?
157	*Faye:*	OK. If the difference between the numbers you have to begin with . . . um . . . is, if you times that by, if you multiply that by itself it will make the difference between the two answers that you get. Yep?
158	*Hazel:*	You agree?
159	*Donna:*	Yeh.
160	*Hazel:*	Right, well done.

The FTA inherent in the request for 'the pattern' (156) is softened by the modal form ('Can you?') and the hedged performative ('sort of explain'). Faye obliges with an account (157) which is characteristically extended, indicating a confident exposition of secure knowledge. It is hesitant only because she begins by reifying a variable – 'the difference between the numbers you have to begin with' – but she must then cope with the burden of the English language, as opposed to symbolic algebra, to say what she wants to say about a function of that variable.

On a number of occasions (e.g. IRG3:156 and notably 83), Hazel asks 'why?' and requests 'explain' with regard to generalities. For the most part these questions and requests seem to require *descriptions* of regularities rather than fundamental accounts of their causes. These two girls might well respond intellectually to some generic example or geometric model of the situation in Episode 1. But, taken as a whole, the transcript is pure delight, a fine example of quasi-empirical teaching and learning, supported by a skilful and sensitive teacher.

Case 2: Ann

Ann is an experienced primary school teacher. In the transcript (IRG2) she is talking to Charlie, a 10-year-old boy in her class. Charlie's mathematical attainment is judged by Ann to be below average for his age.

The conversation is essentially an exploration of residue classes modulo *n*. Charlie will share sweets among people to see how many are left over. Ann has prepared a 'table' on squared paper: she has written 2, 3 . . . 12 along the top (for sweets) and down the left-hand edge (for people).

Ann begins by asking Charlie to share two, three, four, five sweets among two people (herself and Charlie), and explains how to enter the number 'left over' in the table. Her instructions are indirect speech acts, presented in modal interrogatory form:

> 9 *Ann:* Can you put one in that square, because you have three sweets and two people, so you have one left over. Can you share these out now? Four sweets. Are there any left over?
> 10 *Charlie:* No.

Charlie has done the 'practical work' sharing out up to five sweets, and has entered 0101 in the first row. He spontaneously predicts the next two. (Perhaps he lacks his teacher's patience?)

> 16 *Charlie:* This is probably going to be nought, one, nought, one, nought, one.
> 17 *Ann:* Think it is?
> 18 *Charlie:* Yes.

He shields his prediction (16); Ann picks up the hedge and asks how confident he is – her 'think' (17) is more root than epistemic. In fact, Charlie is sufficiently confident (20, 22) to extend the table without recourse to further sharing of sweets:

> 19 *Ann:* Would you like to put it in before I give any more sweets out?
> 20 *Charlie:* Yes.

Evidently Charlie would 'like to', but Ann is uneasy:

> 21 *Ann:* Are you that confident?
> 22 *Charlie:* Yes.
> 23 *Ann:* Go on, then. Put them in. [Pause for Charlie to put in the numbers]

Ann's next instruction is thinly veiled by conventional indirectness:

> 23 *Ann:* Shall we check to make sure?

The 'we' (23) expresses the teacher's solidarity with the child in the activity, and compounds the impossibility of Charlie's refusing the 'invitation' to 'make sure'.

Refusal is clearly dispreferred and would offend the teacher's face wants – both positive (she is a partner in the activity and is suggesting a check) and negative (she is the teacher, he the pupil). Charlie checks his prediction, and Ann moves on to three people:

23	*Ann:*	This time there are three people. Three people. If I get two sweets, can they have one each?
24	*Charlie:*	Yes.
25	*Ann:*	There are three of us, remember.
26	*Charlie:*	I mean, no.

These four turns, frozen in text, illustrate how transcripts can be used to reflect on classroom interaction and to develop questioning styles. Ann's question (23) is presented as a binary option – 'can they?', yes or no? Her response to his answer is neither confirming nor neutral, and leaves him in no doubt that he is being corrected. Naturally, he changes his answer. There is no evidence in the transcript as to whether or not he changes 'his mind'.

Ann, who donated the transcript, might use it to consider alternative ways to IRG3:23 of formulating the question. A number of possibilities come to mind. Why the 'I', which requires Charlie to put himself in her position? Who are the three people (a ruler is adopted later)? But in particular, it may be more effective in achieving intellectual participation from Charlie if it is presented to him not as a two-way choice, but in a more open ('Wh – ?') form such as 'What would happen if you . . .', or indirect variants such as 'What do you think would happen if you . . .'. Such a form would also be likely to elicit a more extended reply, with greater potential for insights into Charlie's personal construction of the situation. No entries are made in the table below the leading diagonal (i.e. one sweet each, none over), yet it is worth considering the fact that there are sweets 'left over' when there are fewer sweets than people, and that the corresponding entries would maintain the cyclic regularity of each row. The question (23) 'Can they have one each?' focuses on the quotient (which is irrelevant to the modular regularity) rather than the remainder.

Under Ann's guidance, then, Charlie enters 0, 1, 2 in the row for three people. Again he spontaneously suggests (with the same Shield, 'probably') an extrapolative extension of the data, but this time he seems to be caught up in the rhythm of the language more than the logic of the situation:

| 38 | *Charlie:* | Probably going to go three, four, five, six, seven, eight, nine, ten. |

This time Ann patiently encourages him (with the conspiratorial 'Let's') to test his prediction:

39	*Ann:*	Let's try it and see. [Counted out six sweets] How many left over?
40	*Charlie:*	None.
41	*Ann:*	[Counted out seven sweets]

42	*Charlie:*	One.
43	*Ann:*	Fine. [Counted out eight sweets]
44	*Charlie:*	Two.

Now she invites him to generalize:

| 45 | *Ann:* | Right. Now can you tell me how the numbers are going to go? |
| 46 | *Charlie:* | Nought, one, two, nought, one, two, nought, one, two. |

The practical work with the sweets is emphasized (laboured?) as (47–50) Charlie is directly instructed to 'Share out and see' for nine sweets. He counts out the ten sweets without attempting a prediction – suspecting, perhaps, that he will have to do it in the end anyway. The only sense-making that is legitimized is in terms of the algorithmic manipulation (sharing) of the embodiment.

47	*Ann:*	What's the next number going to be?
48	*Charlie:*	Nought.
49	*Ann:*	Share out and see. Well done! What about the next number? [Pause while Charlie counts out sweets]
50	*Charlie:*	One.

There is evidence in the transcript that, despite the practical work, Ann is directing Charlie's thinking towards the patterns of numbers on the table rather than the sweet-sharing. Is the 'train spotter's paradise' in sight (Hewitt, 1992)?

| 55 | *Ann:* | Right. Look at that and can you tell me what is going to happen on the next line? [Sharing between four people] |
| 56 | *Charlie:* | Mmm. I think we will get a three instead of a nought. [He is apparently referring to the second zero in 012012, since he writes 012301230 below] |

And again, later:

63	*Ann:*	We have five people now. Where are you going to start on our table?
64	*Charlie:*	There [at (5, 5)].
65	*Ann:*	What do you think the pattern is going to be in the numbers?
66	*Charlie:*	Two, three, one, nought, two, nought, one, I think.
67	*Ann:*	Write it down there to remind you. Why do you think it's going to be that?
68	*Charlie:*	Could be nought, one, two, three, four.
69	*Ann:*	Put that number down as well. So could be nought one two three four? Let's try and see.

Charlie hedges (66) or modalizes (68) his pattern predictions, which are still insecure (66). Ann's enquiry in IRG3:67 is the first of only two 'why?' questions in the whole transcript, and this one causes Charlie to self-correct.

Ann's use of pronouns frequently associates her with Charlie's progress in quite a personal way:

77 *Ann:*   All right. You do that line **for me**. Well done.

79 *Ann:*   **We** put a nought when there is nothing left over, didn't **we**?

The 'we' traps Charlie into complicity with Ann. At the very end, Ann assesses, with further reference to 'me' and now 'us', whether Charlie has understood – or at least remembered.

88 *Ann:*   So can you tell **me** what the nought means? What is it telling **us**?
89 *Charlie:*   That, um, you can share them out and have the right amount.

I suggest that Charlie's 'you' is not, however, addressing Ann (recall, from Chapter 5, that pupils rarely address their teachers directly in mathematical discourse), but an indication that he is articulating a generalization, that he can be detached about the significance of the zeros.

The general purpose of the activity is very nice, very significant – the least positive remainder is less than the divisor. In effect, Charlie has been investigating the Division Algorithm: $\forall a, b \varepsilon N \, \exists q, r \varepsilon N$ s.t. $0 \leq r < b$ and $a = bq + r$. On the evidence of the transcript, Ann has accurately predicted that Charlie will need a good deal of guidance and practical support as he works through the investigation. The same evidence suggests that Ann's attention is on his practical performance rather than his cognitive structuring and his propositional attitude – the way he construes the patterns in relation to the sweet-sharing and the strength or fragility of his conviction. There is no sign of examination of conjectures in the ZCN. She seems to find it hard to release him from the practical task, from dependence on her as teacher and even from obligation to her as 'partner'.

Case 3: Judith

Judith is a newly qualified teacher in an 11–16 secondary school. In the transcript she is talking to Allan, an 'average' Year 9 (age 13–14) pupil.

The conversation (which must have lasted about forty-five minutes) concerns the problem of drawing line segments between pairs of dots in a plane array of some sort (usually regular). What is the least number of segments necessary so that they form a continuous line connecting all the points? It falls into five episodes:

- Episode 1. Investigation of a 3×3 square array (IRG5:1–49).
- Episode 2. Prediction and verification for a 4×4 array: expansive generalization for square arrays to include 3×3 as a special case (IRG5:50–93).
- Episode 3. Prediction, verification and expansive generalization for rectangular arrays, including squares as special cases (IRG5:94–151).

- Episode 4. Consideration of triangular arrays: reconstructive generalization (Harel and Tall, 1991) based on number of dots in the array which includes the three generalizations arrived at inductively in the previous episodes as special cases (IRG5:151–160).
- Episode 5. Search for regularity in the sequence (of numbers of line segments) for triangular arrays – the arithmetic sequence of differences (IRG5:160–264).

The episodic overview immediately reveals the richness and cognitive complexity of the layers of generalization that are built up in the first four episodes. The first of these begins:

> 1 *Judith:* OK, Now. Here is the investigation, so you can read it. Draw a three-by-three dot grid. Start anywhere you like. Draw a continuous line that goes to every dot. Yeah?

Notice that we have entered the culture of the post-Cockcroft secondary school – 'doing an investigation' (Love, 1988, p. 250). In contrast to the examples from primary schools, the investigation is presented to Allan in a written format. This may reflect Judith's inexperience (it feels 'safer'), but in any case it is how investigation 'starters' *are* normally presented to secondary school pupils, for GCSE coursework and the like.

Judith makes reference to the rituals of 'doing an investigation' culture on several occasions in the interview:

> 88 *Judith:* So if you were doing an investigation what would you write down for me?
>
> 160 *Judith:* So... what would you write down if you were doing an investigation?
>
> 130 *Judith:* ... if you did it three, four, five, six it might be easier to see patterns, do you not think? Do you do tables when you do investigations...?

Judith exploits the familiarity of the investigation 'write up' to encourage Allan (88) to articulate his thoughts, and with some success:

> 89 *Allan:* I'd write... that the pattern is... if you ti—, times both, if you square the side, the side, and um, you minus one, you'd be th', the amount of dots you—, if you went round the dots it'd be the same answer.

Allan's approach to speech as imagined writing seems to assist him in assembling his thoughts, and he gives (with a few false starts) a relatively formal account of a generalization. Presumably oral 'reporting back' (Pimm, 1992, pp. 68–72) is not a part of the practice of school investigations with which Judith expects Allan to be familiar, since it would otherwise be a more natural point of reference for him. Writing is, on the whole, more dense, more formally structured than speech (Perera,

1990; Brown and Yule, 1983, p. 15), and Allan still has need of the informal generalizer 'you' to formulate his rule.

To return to Episode 1: Allan is quite relaxed and competent in the use of epistemic hedges:

11	*Judith:*	OK. Do you think you're going to be able to do it in less than . . . that? . . . nine?
12	*Allan:*	Maybe, yeah.
13	*Judith:*	Maybe.
14	*Allan:*	Maybe.
15	*Judith:*	OK, so you're not . . .
16	*Allan:*	Not positive, but I am . . .
17	*Judith:*	OK [pause]. Top right [pause]. Are you thinking about where to go next?

Judith gently echoes (13) and explores (15) his uncertainty; her interruption (17) relieves the tension but cuts off the flow of data from the informant. Before long he has 'done it' with eight line segments (starting with a corner dot), but he remains tentative (24, 30):

24	*Allan:*	Still eight again, so probably the most is eight.
25	*Judith:*	You mean the least?
26	*Allan:*	Yeah, the least, sorry.
27	*Judith:*	That's all right.
28	*Allan:*	Yeah.
29	*Judith:*	OK, so do you think starting not in the corner could get you eight as well?
30	*Allan:*	Um, possibly, yeah. [Pause]

His doubt appears at first to be well founded:

40	*Allan:*	So it don't work from the middle at all, really, because it's, uh, because you have to go in and out again.
41	*Judith:*	D'you think? Are you sure?
42	*Allan:*	Because what you have to do, you have to go, from the middle you have to go all the way round and you go into one and come back out again, but if you do it from, like we 'ave did from exactly, exactly in the middle.

Allan's utterance (40) provides examples of procedural deixis ('it don't work') and the generalizer 'you'. His 'really' (an adverb he uses again in IRG5:56 and 217) is a hedge on his claim that 'it don't work', much the same as 'basically' (Chapter 6). Judith tests his claim (41) in the ZCN, and Allan responds with an extended if somewhat incoherent account, making heavy use of the generalizer 'you'. I am reminded of one particular moment when Susie 'explained' to me with a long, rambling and quite incomprehensible speech (S1:30) to which, completely lost, I weakly responded, 'So what did you do next?' in the hope of being offered some guide to interpretation. Judith's next turn (43) has some of that 'lost' quality:

43 *Judith:* Try that, then.
           [Pause. Allan draws a route from the middle]

In fact, Allan finds the counter-example to his own argument:

44 *Allan:* Yeah, made eight as well from the middle.

Judith now needs to know quite where Allan stands at this stage. She uses the 'if you were doing an investigation' strategy to ask for a summary progress report:

45 *Judith:* OK. So what do you think? [Pause] So if I wasn't talking to you and you were just doing the investigation yourself, what would you be thinking?
46 *Allan:* I'd be thinking that the most possible, um, way, uh, of getting, of getting, getting the least is that it only started from the middle going right round, going right round or going, going from one corner but you can't do it in the middle, middle between two corners.
47 *Judith:* OK.
48 *Allan:* And also the least is eight.

She invites Allan (ushering in Episode 2) to determine where the investigation will go next:

49 *Judith:* All right then, so what're you going to do now?
50 *Allan:* I'll try a, um, four-by-four grid.

Once he has decided, she explores – with an indirect request (51) for a prediction – how his thinking is becoming structured, whether he has any general overview of the problem.

51 *Judith:* Right. Can you make any predictions before you start?

The indirectness softens the force of the FTA, and his hedged prediction (54) suggests that he may have formed a generalization from the single 3×3 instance.

54 *Allan:* The maximum will probably be, er, the least'll probably be 'bout fifteen.

The epistemic adverbial Shield 'probably' is reinforced with the Approximator hedge '(a)bout'. This is an interesting use of an epistemic Rounder (Chapter 6), repeated later by Allan (below) in (73). Incidentally, it is plausible, but unlikely in my view, that 'fifteen' is being used as a 'round number' (Channell, 1994, pp. 87–89) and an Approximator in its own right. Judith wants to explore the thinking behind the prediction:

68   *Judith:*   So why did you predict fifteen?
69   *Allan:*   Uh, because I thought there might be a pattern between . . . if there
                 was, um, a certain amount of, um . . . if it's three by three, say . . .
70   *Judith:*   Uh-hum.
71   *Allan:*   If you ti—, three times three is actually nine.
72   *Judith:*   Uh-hum.

To begin with, Allan is struggling; perhaps the linguistic struggle is the manifestation of a meta-cognitive struggling to recover or construct the reason for his prediction. His overture ('I thought there might be') is uncommitted. At first he finds it helpful to illustrate (69, 71) with the 3×3 example, rather than to articulate the generality. Judith waits; her interventions (70, 72) are absolutely minimal. She says just enough to assure Allan that she is listening, like a counsellor listening to a client. She is soon rewarded with an explanatory outpouring, punctuated with the impersonal 'you':

73   *Allan:*   But as, if you went round all the dots, it would only come to about,
                 if you did it once it would come to one, uh, less than nine, 'n' you
                 got, uh, because, because there's o—, there's only . . . cause you
                 only have, y— . . . you can miss out a line exactly, 'cause y-you can
                 miss out a gap, c-'cause you um, y'd 'ave to go all the way round
                 the whole dots.

So what, asks Judith (74), are the significant generalizing features of this 3×3 example? What does the generalization look like (76)?

74   *Judith:*   OK . . . So why did that make you say fifteen?
75   *Allan:*   Because uh, f— for the same reason, 'cause if you, um, w— tried to
                 go round the whole, all the dots you'd get sixteen but if you just did
                 it once all the way round the dots but missing out gaps you'd still
                 come to, uh, you just minus one basically and just . . .
76   *Judith:*   So what would happen in some other squares?
77   *Allan:*   Probably if you minus one from the s—, if you square the number
                 you'd probably find that if it was actually, if you minus one from
                 that you'd probably find that that would be the answer to the . . .

The most refined 'algebraic' account that Allan is able to develop (77) is divorced from the dots and the gaps. It is an algorithm, albeit a highly ('probably') tentative one. I'm still here, says Judith (78):

78   *Judith:*   OK.
79   *Allan:*   . . . to how many dots there are, to how many times you 'ave to go
                 round the dots.
80   *Judith:*   OK. Now do you want to try one, or are you certain of that?

Judith's question (80) seems to present Allan with a genuine option. It really is a genuine question rather than an indirect instruction. The 'or' is explicit. There is no obvious preferred response. Contrast it with Ann's earlier (case 2):

> IRG2:23   *Ann:*   Shall we check to make sure?

In fact, Allan goes on to say that he is not certain, and tries out the 5×5 array.

I have examined less than one-third of the transcript; fifteen minutes of talk perhaps. My purpose in studying these transcripts was to look in them for the linguistic features that I had identified in my own conversations with pupils, and for some validation of the conclusions I had reached about the pragmatic function of vague language. It was not my intention to say what was 'good' or 'bad' about the teaching, but the analysis inevitably puts into sharp relief those aspects of practice which support or negate a conjecturing atmosphere – assuming that that is what is wanted. In that respect, Judith's instincts are remarkably true.

Case 4: Rachel

Rachel is in her third year of teaching. She works with 16 to 18-year-old students at a sixth-form college with a strong academic reputation. Rachel interviewed two pairs of 18-year-old students, all following an Advanced Supplementary (AS) course in mathematics.[1]

The students were presented with the 'stairs' investigation (see Chapter 2) in the following – written – form.

> You need to climb a staircase with n steps. You are allowed to go up the steps, taking either one or two steps at a time. In how many ways can you go up n steps? As an extension consider being able to take one, two or three steps.

In the first transcript supplied by Rachel (IRG6A), the two informants, Juliette and Di, are described as bright students who work well together in class. In the event, both gained A grades in the AS examination. Rachel begins by checking that the task is clear.

> IRG6A:1   *Rachel:*   So, do you think you understand what it means?

Juliette clarifies the task, seeking confirmation with the tag question 'can't you?' (3, 7)

> 3   *Juliette:*   So you can have a combination of ones and twos, can't you?
> 4   *Di:*   It's going to take n steps or n over two steps.
> 5   *Juliette:*   Or a combination. You could look at . . .
> 6   *Di:*   You could take n as being one [writing a table], x, y.

> 7   *Juliette:*   If *n* is two, you could do either two steps or one, can't you? *n* equals three . . . [writing down 2 1, 1 2] Does it count if you do two and one and one and two?
>
> 8   *Rachel:*   Yes, they're different.

In IRG6A;3, 5, and throughout the interview, Juliette freely uses 'you' for generalization and/or detachment. Di quickly goes into 'investigation mode' with a table of (x, y) values. Juliette clarifies the rules on 'sameness' ('Does it count . . . ?') and Rachel adjudicates (8). 'Does it count?' seems to be a standard legitimation enquiry, just as 'It works' is a standard procedural generalization. By IRG6A:10 Di has made a (false) prediction of four ways for four steps:

> 10   *Di:*   Yeah, [writing out an (x, y) table] so you have one, one; two, two; three, three; so four is four.

Her prediction is not marked in any way as regards uncertainty. What is the interviewer – also, here, the teacher – to do?

> 11   *Rachel:*   I think maybe you need a few more before you can generalize.

The double plausibility Shield (11) is, of course, play-acting on Rachel's part. Rachel wrote a reflective account of her two interviews after she had transcribed them (Williams, 1995) and comments:

> I feel I have to interrupt and prompt her to consider a few more cases. In retrospect, perhaps I should have waited to see if Juliette did that.

Soon, the two students are working on the case $n = 5$:

> 18   *Juliette:*   . . . Two, one, one, one, one, two, one, one, one, one, two, one. [More writing] That's it. So, we've got one, two, three, five, eight and now you're going to get twelve.
>
> 19   *Rachel:*   Why?
>
> 20   *Juliette:*   It's a series. You add one. You add one, that's, that's . . . you add one. Oh, I don't know.
>
> 21   *Rachel:*   Yes, yes, you add one.
>
> 22   *Juliette:*   You add one, then you add two, then you add four, the interval between. No, that's not right. It's something to do with the series it goes up with.
>
> 23   *Rachel:*   Mmm.

This time it is Juliette who is misled to a faulty prediction (twelve ways for five steps) by an alternative regularity (22) in the first few terms. Again, her prediction (18) is unhedged. She freely decentres with 'you' (18, 20, 22). Note that 'we've' got one, two, three, five, etc. (actual data), whereas 'you're' going to get twelve (prediction). Finally 'I' don't know (a personal epistemic state). Rachel asks for an explanation (19), which is taken to be a request for an account of the perceived regularity (20). Rachel enthusiastically encourages (21) Juliette's faltering start, which seems to be based on addition. The continuation (22) perpetuates the faulty prediction, but is not committed to it. Rachel's response (23) is minimal (cf. Judith, IRG5:70, 72]. Later she reflected (Williams, 1995):

> Juliette has the right sort of idea [22] and says 'It's something to do with . . .' but she will not commit herself. I am trying hard not to interfere and biting my tongue with 'Mmm'.

In the end Di is surprised to find more ways than Juliette had predicted (31) and Rachel can restrain herself no longer:

31	*Di:*	[counting all the combinations] Ten, eleven, twelve, thirteen. [Delighted/puzzled]
32	*Rachel:*	That's all right.
33	*Juliette:*	[puzzled] That's OK?

Rachel's comment:

> I had to confirm that she was correct [32], I couldn't bear the uncertainty and wanted them to know they had got to the correct number of ways. Looking back, it would have been better to let them sort it out. (*ibid.*)

Rachel brings out another affective dimension in the conjecturing atmosphere. The pupil is required to take risks, but the teacher may have to 'bear the uncertainty' when she judges that the pupil must resolve it him/herself. That is not to say that the teacher cannot participate in the ZCN, but her/his role may be best restricted to light, indirect, linguistic scaffolding.

The two students are hooked on familiar sequences (arithmetic, geometric) and are somewhat inflexible in their search for regularity in the (Fibonacci) sequence 1, 2, 3, 5, 8, 13 . . . In the end, Rachel finds it hard to allow them to flounder:

35	*Juliette:*	Times two, no, no direct . . . no common difference, no common ratio.
36	*Rachel:*	But when you were doing your thing of adding on each time it worked. Well . . .

Rachel later commented:

> I sense that I am hesitant in letting them search too long for a solution and that I assume they want to get to the answer quickly. I am in some way anxious that it is taking them a long time, but in interrupting, I interrupt interpreting conversations they could have in getting to the answer . . .

> I was anxious that it should work and I think that bright students should be able to get to the answer without too much difficulty. I did have expectations of the students . . .

I felt that their knowledge got in the way of their intuition. They felt that the question should fit into an arithmetic or geometric progression type question.

I felt that I intervened too quickly and didn't let them struggle enough. Perhaps I was more nervous of the tape than they were – this was my first interview. (*ibid.*)

In her second interview (IRG6B), Rachel discussed the same problem with Clare and John. Rachel judged Clare to be the more able of the two students; in fact Clare later gained an A grade in the AS examination, John a C. She takes the lead in the conversation, and soon arrives at and articulates the Fibonacci 'rule'.

> 14 *Clare:* So that gives us two, three, five, eight. You just add on the previous number.

Prompted by Rachel, she predicts and verifies the next term, 13.

> 15 *Rachel:* Do you want to check it, then? What would you predict the next one would be?
> 16 *Clare:* Hmm, thirteen.

There is no evidence of an attempt to account for the observed and confirmed regularity, and no enquiry as to John's commitment to it. They move on to expansive generalization:

> 19 *Rachel:* OK. So try this one, then. If you could take one or two or three steps.

Before long, Clare articulates and revises a provisional generalization, based on the first three terms of the sequence:

> 25 *Clare:* Do the next one then, four. It's squared, so it'll be nine. Probably. No, that's double two, you double then . . .

Clare is resourceful in generating conjectures for 'the pattern' on the basis of the data available; thus, with the terms 1, 2, 4, 7 she ventures:

> 30 *Clare:* All right, then, you add up all the numbers before.
> 31 *John:* Seven.
> 32 *Clare:* So the next one will be fourteen, maybe? Three, seven, fourteen. Do you think maybe you just add – mm, no, it doesn't work there. OK.

The provisional status of the generalization (30) only becomes apparent in the Shields (32) that she uses when she applies the rule to predict the next term $(1 + 2 + 4 + 7 = 14)$. Very soon she modifies the rule – it is not clear from the transcript whether they find that there are in fact thirteen ways for five steps:

> 36   *Clare:*   Add the previous three, because when you had two steps
> you added the previous two numbers and then you have
> three steps, you add up the previous three numbers.

This generalization is expansive in so far as Clare relates it to the 'rule' for 'two steps'. The 'because' refers to the *form* of the generalization rather than to any underlying reason. There is no linguistic sign of uncertainty at this point.

Perhaps because of Rachel's recognition of the face wants of her students, her desire for and expectation of their success, she seems to have difficulty in allowing them to struggle. There is a hint of bravado in the three female students' approach to the problem, and little evidence of uncertainty. Perhaps, without being pressed to consider proving their inductive conjectures, the problem is not in fact much of a threat to their mathematical self-esteem. John's contribution to the conversation is marginal, dealing with minor clerical and arithmetic matters. His turns are brief, and give little insight into his propositional attitude.

Case 5: Sue

Sue, an experienced teacher, works with the reception class in a primary school. She provided transcripts (IRG1A–D) of four short interviews with 4-year-old children: A: Rebecca, aged 4:8 (4 years and 8 months); B: Jane, aged 4:8; C: Anna, aged 4:3; D: Jason, aged 4:11. The task in each case was as follows. Sue placed five plastic 'people' on the floor, and asked the child how many people there were. Next, she asked the child how many more people would be needed to make ten. In two cases she then asked how many more would be needed to make twenty.

I shall consider the interviews together with reference to some common features, some of which relate to my observations about Year R children (under 'Initiation') in Chapter 7.

Naivety and Directness

Adult behaviour and questions are taken at face value, the child is naively co-operative, and simply acknowledges her/his ignorance (D10), error (C4) or uncertainty (C8, 10) for what it is:

> D9   *Sue:*   How many more do we need to make twenty?
> D10   *Jason:*   Umm, don't know.

> C2   *Anna:*   [pause] Umm, ahh, one, two, three, four.
> C3   *Sue:*   Four, you think?
> C4   *Anna:*   No, one, two, three, four, five.

> C8   *Anna:*   One, two, three, four, five, six, seven, eight, umm, nine, I think we
> need nine more.

C9   *Sue:*   You need nine more?
C10   *Anna:*   I think so.

Like Jason, Jane knows her limitations, and it does not appear (B10) to be a face-threatening issue for her.

B9   *Sue:*   How many more do you think we'd need to make twenty?
B10   *Jane:*   I don't know. I can't count up to twenty. Only my sister can.

Teacher Language

Nevertheless, Sue is frequently indirect in her instructions to the children. In the later meetings of the IRG it was very apparent that Sue wanted to create and sustain a conjecturing atmosphere in her classroom, one in which the children knew it is 'all right to be wrong'. Thus her first question is direct in every case, presumably because it is the bread-and-butter of the reception class:

C1   *Sue:*   Right, Anna, how many people are on the floor?

whereas the next question, which she suspects (with good cause) will challenge them, is twice endowed with a Shield, and always with 'we' for solidarity:

C5   *Sue:*   How many more people do you think we need to make ten people?

Her first question (C1 and see B1 and D1 below) is in every case heralded with an utterance initiator, 'Right' (three times) or 'OK' (once). This interests me because I know (from transcripts) that I use 'right' a great deal. On reflection, I suggest that 'right' and 'OK' have three distinct pedagogic functions. The first (that identified above) is to indicate boundary points in a 'lesson'; such markers (which also include 'now', 'now then' and 'well') are called 'frames' by Sinclair and Coulthard (1975, p. 22), who observe that they initiate a teaching 'unit' in which they are typically followed by a meta-statement of some kind. For example:

I:8.1   Right, we're going to start with a quiz today.

The second function is a minimal interjection into a pupil's account of something (an explanation, for example) or attempt to formulate such an account. It assures the main speaker (the pupil) of the listener's attention. See particularly case 3 and case 8 for examples. The third function, as a statement tag, is to seek approval (pupil use) or to seek assurance of agreement or comprehension.[2]

In the two cases where the child incorrectly assesses the size of the initial set, Sue counters with the Attribution Shield 'you think', which I have noted (Chapter 6) as a feature of my teacher strategy. For example (to Rebecca):

A7   *Sue:*   You think six. How many more people do you think we would
                need to make ten?

Counting

It is very clear that (see Chapter 7), for three of these four Year R children, a 'How many?' question triggers a count. The reaction is almost Pavlovian:

B1 *Sue:*   Right, Jane, how many people are there?
B2 *Jane:*   One, two, three, four, five.

D1 *Sue:*   OK. Jason, how many people have we got here?
D2 *Jason:* One, two, three, four, five.

Examples of the same phenomenon are recorded elsewhere (e.g. Walkerdine, 1988, p. 106). Recall that 'when you ask children "How many . . . ?", they count' (Nunes, 1996, p. 74). The response to the stimulus question 'How many . . .' is a recitation out loud of the 'standard number word sequence' (SNWS – Steffe et al., 1983, p. 25) of 'numerons' or counting 'tags', but there is no external evidence that the child has mastered the cardinal principle, that the last tag used is the cardinality of the set, in evidence of which

> [. . .] the child must be able to pull out the last numeron assigned and indicate that
> it represents the numerosity of the array. (Gelman and Gallistel, 1978, p. 80)

There may, of course, be prosodic features of the children's counts that 'pull out' the last number in the recitation, but this is not indicated in the transcripts.

On 'more'

Lack of mastery of the cardinal principle may be related to the non-standard response of Jane and Anna to the 'How many more?' question.

B3 *Sue:*   How many more people will we need to make ten?
B4 *Jane:*   [pause] Umm, six, seven, eight, nine, ten.
B5 *Sue:*   Right, do you think you could find some people to make ten?
B6 *Jane:*   Just yellow?
B7 *Sue:*   I don't mind. You can have whatever colour you like.
B8 *Jane:*   One, two, three, four, five, six, seven, eight [pause to rummage] and
            a yellow one [puts down two more] ten.

Jane's response (B4) is to extend the recitation of the SNWS beyond the point she had reached ('five') to 'ten'. The purpose of Sue's contingent question (B5) seems to be to discover whether Jane will match the five numerons ('six' to 'ten') with five people. In fact Jane legitimately interprets B5 as a request to get a new set of people to 'make ten'.

  Anna does offer a prediction – unhedged – (C6), adds about the right number to the original set of people, but suggests (Shielded) in C8 that the cardinality of the union is answer to the 'more' question.

C5	*Sue:*	How many more people do you think we need to make ten people?
C6	*Anna:*	Three.
C7	*Sue:*	Three? Do you want to find out? [Anna adds more people] Now how many have you got now?
C8	*Anna:*	One, two, three, four, five, six, seven, eight, umm, nine. I think we need nine more.

Similarly, Jason's response to 'How many more do we need to make twenty?' (IRG1D) was to count (with an unstable SNWS) to twenty.

Despite Sue's experience as a teacher of young children, these responses surprised her. She wrote (personal communication):

I found I had made assumptions about their basic mathematical language. We have taken a great deal of trouble (we use the Ginn[3] reception maths teacher's book and ideas of our own) to teach the children what they will need [. . .] but many did not understand 'more', for example.

The problem probably originates in the particular precise and situated meaning of 'more' and 'How many more?' in the social practice of school arithmetic (Walkerdine, 1988, pp. 22–27). Walkerdine notes that, in the child's home, 'more' is associated with 'food regulatory practices' (p. 26), so that (for example) the opposite of 'more' is not 'less' but something like 'no more' or 'enough'.

Rebecca is something of an exception in these four interviews. She is alone in not being kick-started into a vocal count by the initial 'How many?' question. She either estimates or internalizes the SNWS (the transcript does not reveal which):

| A1 | *Sue:* | Right Rebecca, how many yellow people are there here? |
| A2 | *Rebecca:* | Umm, six. |

There is evidence that this is in fact an estimate, in that she twice (subsequently) asks Sue whether she should count (A6, 10) and, demonstrates, moreover, that she can count small sets accurately. She appears also to estimate how many more are needed to make ten.

| A4 | *Rebecca:* | Umm, two more. |
| A5 | *Sue:* | Two more? Do you want to try that, then? |

Sue is correcting Rebecca by repetition of her answer (Drew, 1981, p. 252), inviting her self-correction. The transcript notes that Rebecca selects three people. Sue continues:

A5	*Sue:*	Right, how many people are there now?
A6	*Rebecca:*	Ten. Shall I count them and see?
A7	*Sue:*	That's a good idea.
A8	*Rebecca:*	One, two, three, four, five, six, seven . . . [pause] Shall I do some more along there?

Rebecca seems to make more conventional sense of the 'How many more?' question. On the other hand, Sue does not press her for an answer to the question.

>   A10   *Rebecca:*   Shall I count them and see again? [Sue smiles] One, two, three,
>   four, five, six, seven, eight, nine and ten.
>   A11   *Sue:*   Thank you, Rebecca.

In later meetings of the IRG Sue commented on her growing awareness of the need to use language which emphasizes the autonomy of the young children whom she teaches. In effect, she wants to equip them with 'basic' arithmetic knowledge and competence, but wants them to retain responsibility for sense-making and validating their solutions to problems. Yet sense-making in school mathematics is not solely a matter of private interpretation within some absolute, secure reality of 'real' objects ('people' and the like); it is also one of of linguistic enculturation, of initiation to a discursive practice (Walkerdine, 1988, p. 128).

Case 6: The Public Lecture

This case is not so much a conversation as an example of a ritual mathematical monologue – the public lecture. I propose merely to set the scene, to present some of the data and typographically highlight some relevant features.

   The lecturer is a mathematician from the United States, elected a 'visiting scholar' by one of the colleges of the University of Cambridge. In return for a year's fellowship the Fellow must deliver one public lecture 'for a general audience'.

   The title of this lecture, given in 1994, was: 'How many lattice points lie in a circle?'. Essentially, the area of a convex region of the plane is roughly equal to the number of lattice points (of a square grid) which lie inside it. Therefore a first approximation to the answer to the question (the lecture title) is πR^2 where R is the radius of the circle. If the error $\varepsilon(R)$ in this approximation is of the order of R^θ for some θ (so that $\varepsilon(R) = kR^\theta$), what bounds can be put on the exponent θ?

   The data which follow were all spoken by the lecturer. From the preamble:

> At the Isaac Newton Institute in July last year, Andrew Wiles reported on his **almost-proof** of Fermat's Last Theorem.
> Surface tension **sort of** holds the water drop together.
> **It turns out that** the differential equation is the same – a form of **hand-waving**
> – I could tell you but . . .

Into the substance of the lecture:

> From numerical studies, a sample with R < 1800 **suggests that** $R^{1.2}$ is better than $R^{2/3}$.
> So this **suggests that** values of R **around** 10^8 are needed to show that $\theta < 0.6$.

Hedging is most apparent in the speaker's necessarily unprepared answers to questions after the lecture proper:

> Half is still a lower bound, but **it isn't so apparent that** seven-elevenths (due to Iwanec and Mazzecci, 1988) is still an upper bound.
> The envelope **suggests that** $\theta < 0.575$.
> **I don't think that** he [G. H. Hardy, 1915] got the conjecture from looking at Bessel functions.
> **I think** that's what led to the conjecture.
> **I don't think** that's been observed yet.
> **We might be able to show that**, by accumulating jumps, we could make the bound rigorous.

Here is a mathematician expert in his chosen field and in the register of the discipline. The significance of the vague language he uses here is that it is perceived not as a deficiency, either in language or comprehension, but as an acquired expertise, enabling him in each assertion to be as precise as he chooses.

Case 7: Open University Video

Debbie teaches in a primary school in Suffolk. She was one of a number of teachers filmed by the Open University[4] production team for course EM236. By a nice coincidence, Debbie was also one of the seven teachers whom I had observed some years earlier in a study of the introduction of CAN (a calculator-aware component of the PrIME project) in Suffolk in 1986–1987. At that time she was a newly qualified teacher. When I interviewed her for the study, she considered that the investigative approach to teaching and learning, which was central to CAN, 'fits in comfortably with what I was doing before' (Rowland, 1994, p. 34), although she recognized that she was giving children more thinking time, 'waiting, prompting if necessary, but not telling them' (p. 35).

In the video Debbie is working with a group of six children whose age (unspecified) appears to be about 7 or 8. The children are considering the number of unit cubes in 'cross' formations made with cuisenaire rods. Her questioning is frequently but not invariably indirect, for example:

EM1:1   *Debbie:*   What sort of shape would you call that? Kathleen?

    9   *Debbie:*   Could you tell me what the ninth one would look like?

  44   *Debbie:*   Ok, can you tell me how many cubes you'd need, then?

She invites the children to extrapolate the sequence of geometric formations (9) and to compute the numbers of unit cubes in them.

  22   *Debbie:*   The ninth cross you've made, how many little cubes would you need to make that?

> 26 *Debbie:* Michael agrees, right. How many little cubes to make the thirty-
> seventh one?

EM1:22 exemplifies the tension between composing sentences with correct syntax ('How many little cubes would you need to make the ninth cross?') and the desire to present the item which is to be the chief focus of attention of the audience (here, the ninth cross) without delay in the sentence.[5] This point is discussed further in case 8. In that case the use of an anaphoric demonstrative pronoun 'that' becomes necessary; the pronoun is co-referential with the object of the subordinate clause which has been highlighted at the beginning of the sentence.

In Chapter 6, I noted my own tendency to use attribution hedges (such as 'Frances thinks that . . .') to sustain the intellectual involvement of other children at particular moments in the conversation, particularly when one pupil has given 'the answer' to a question. Debbie uses the same linguistic strategy, but more as a device to sidestep evaluation of their answers and suggestions:

> 20 *Debbie:* Now you think that's the ninth one. If the box only had hun-
> dreds of these little white cubes in, how many would you need
> to use to make the second one?

EM1:26 has the same quality, in that Debbie is the chairperson of the discussion, ensuring that the state of play is understood by all, the views of all are heard and considered:

> 38 *Debbie:* Is that the thirty-seventh one, Alex?
> 39 *C3:* It is.
> 40 *Debbie:* Alex doesn't look convinced.

When the children are asked how many unit cubes there would be in the thirty-seventh cross, the girl C5 displays competence in the use of marked language – hedges and modal forms. The cross is an object which they must conceptualize first and construct later as a conservative extrapolation of the smaller, more tangible formations.

> 26 *Debbie:* Michael agrees, right. How many little cubes to make the thirty-
> seventh one?
> 27 *C?:* Oh!
> 28 *C5:* More than a hundred, I should think.

C5 seems to be offering a vague estimate of the kind of size she would expect, and the estimate is appropriately hedged. She gives further emphasis to the hypothetical nature of her estimate with the modal 'should'. The modal-hedge combination 'I should think' is doubly cautious, and unusual for such a young child – it does not occur anywhere in the Make Ten data.

Later the two girls miscalculate the number of unit cubes; C5 indicates her lack of full commitment, again with an epistemic modal (46):

46	*C5:*	It might be . . .
47	*C5–6:*	A hundred and seventy-nine.
48	*Debbie:*	How did you get that?
49	*C5:*	Well . . . [laying out a limb with three 10s] one, two, three . . . [lays a 5] thirty-five [lays two 1s] thirty-six, thirty-seven.

C5's 'Well' in EM1:49 suggests a possible Quality hedge, i.e. they didn't in fact get their (erroneous) answer by laying out the cross in cuisenaire. When they do, C5 counts the tens, then the fives, then the twos, and correctly computes four thirty-sevens.

The nature of the learning here does not involve the children in making inductive conjectures or enthymematic leaps.[6] Debbie is focusing the children's attention on the geometric generalization in the form of the crosses – four arms of equal length, with an additional unit at the centre. C1 gives evidence of having made this generalization.

45	*C1:*	Oohhh! . . . four times thirty-seven, no . . . yeah. Four times thirty-seven.

Well, almost . . .

54	*C2:*	You didn't count the one in the middle.
55	*C5:*	Oh yeah!

Paul Cobb and his collaborators provide a brief but pertinent discussion of the teacher's role in classroom conversations in which he or she participates, appearing inevitably as an authority figure, yet having come to that interaction with commitment to a constructivist theory of knowing.

> One feature of the teacher's active and demanding role is therefore to facilitate mathematical discussions between students while at the same time acting as a participant who can legitimise certain aspects of their mathematical activity and sanction others. In doing so, the teacher ideally provides a running commentary on the students' constructive activities from his or her vantage point as an accultured member of the wider community [. . .] in a communicative context that involves the explicit negotiation of mathematical meanings. (1992, p. 102)

A significant factor in the above teaching sequence (case 7) is Debbie's avoidance of both legitimizing and sanctioning, her refusal to assume the role of truth assessor, insisting rather that the children should take responsibility for the validity of their own solutions, 'which must occur in order to allow the construction of meaning' (Balacheff, 1990, p. 259).

Case 8: Jonathan

My last case study concerns Jonathan; at the time he was one of my undergraduate mathematics/education students. The four-year course that he followed includes substantial academic study of mathematics, and one of his third-year options was my course in the Theory of Numbers. The paper is assessed by a three-hour examination and two twenty-five-hour 'projects'. Students choose their projects from a menu of possible starting points, and are normally given two one-to-one 'supervisions' on each project. I tape-recorded supervisions with a number of students; I have selected one (NT4) with Jonathan for analysis.

Jonathan came to see me at 9.15 on the last Monday morning of the Lent (spring) term. The appointment had been made at the conclusion of our previous supervision meeting a week earlier. For this project, Jonathan had been working on the problem of finding the number of integer solutions of $x^2 + y^2 = n$ modulo a prime, p. At a previous supervision he had discussed the cases $n = 0$, 1 with me, with some conjectures, and a proof concerning the number of solutions when $n = 0$. Since then he had generated some data for $n > 1$.

We talked for about forty-five minutes; the transcript (NT4) can be separated into five episodes:

- *Episode 1*. Recall of the previous supervision, including a sketch of the proof of a theorem about the number of solutions when $n = 0$ (NT4:1–73).
- *Episode 2*. Elaboration of the proof, guided by Tim (NT4:74–135).
- *Episode 3*. Jonathan talks about the case $n \neq 0$ (NT4:136–156).
- *Episode 4*. Discussion of particular values of $n \neq 0$; when n is a quadratic residues mod p; proof that the number of solutions is the same for all such n (NT4: 157–208).
- *Episode 5*. Jonathan's proof that, for all $n \neq 0$, there are $p + 1$ solutions if $p \equiv 3 \bmod 4$, and $p - 1$ solutions otherwise (NT4: 209–240).

Much of the generalizing, the forming of conjectures, had taken place at the previous supervision. The process most to the fore in this encounter is proof. Jonathan's ideas are mostly skeletal, in need of detail, and my role (as I perceive it) is to provide some scaffolding around his construction of the details.

First, I need to ascertain what the conjectures are, and what progress Jonathan has made with the proofs. There is discomfort in the transcript as Jonathan is submitted to cross-examination. Indeed, a hallmark of the whole transcript is hesitation on the part of both speakers; both of us frequently seem to find it difficult to 'spit it out'.

1	*Jonathan:*	Well, I had a bit of a bash this time with the theory.
2	*Tim:*	Right . . . we're talking about x squared plus y squared equals n . . .
3	*Jonathan:*	Yes
4	*Tim:*	Yes, yes . . . um, can you . . . did we discuss equals zero last time?

5	*Jonathan:*	It . . . came up, yes.
6	*Tim:*	Right . . . you'll forgive me, but I've discussed the same/
7	*Jonathan:*	/yeah (yes)/
8	*Tim:*	question with one or two people.
9	*Jonathan:*	Yes, and . . . that . . . that I'm quite ha . . . well, fairly happy with the argument I can put for that one.
10	*Tim:*	[rising pitch] All right.
11	*Jonathan:*	That's, that's, that's the happiest argument that I've got. [Laughs]

My laboured formulation (4) of Jonathan's progress indicates my embarrassment (6) that my recall is uncertain. It is a 'deferential use of hesitation and bumbliness' (Brown and Levinson, 1987, p. 187) which shows my reluctance to reveal to Jonathan that my memory of our last supervision has merged with that of similar conversations with other students. I try to redress the FTA with a request for acquittal (6). Jonathan (9) stumbles over his words as he asserts that has a proof for the case $n = 0$, and hedges his satisfaction ('quite/fairly happy') with his argument.

I urge Jonathan to state the theorem for $n = 0$, and, after two or three attempts, he tells me that there is just one solution ($x = y = 0$), when $p \equiv 3 \mod 4$, and $2p - 1$ solutions when $p \equiv 1 \mod 4$ (NT4:12–27). At this point I wonder whether we'll ever get round to the proof, and I am anxious that Jonathan should be aware of the significance of $p \equiv 1 \mod 4$ relative to the theorem he has enunciated. In retrospect, I should have asked him to elaborate his argument first – once again, 'the tape recorder trains the teacher'. My intervention (28) is an imposition on him, since he has told me he has 'an argument'; reluctance to perform the FTA is evident in the hesitation (28) which has no fewer than six false starts, as I search in vain for a way of not telling him the significance of the value of p modulo 4. The best I can manage is to avoid telling him which case is which!

28	*Tim:*	OK, OK. And, I mean, can I, I think, I just want to ask, does it hinge on the fact that in one case minus one is a quadratic residue and in the other case it isn't?
29	*Jonathan:*	[pause] Um . . . well, yes [coughs] . . . sort of. Um, I mean /it's, yes there's one/
30	*Tim:*	/[laughs] Would/ you like to rehearse the argument with me, or . . .
31	*Jonathan:*	Well [coughs], yeah (yes), I'll come back to that bit about the quadratic residue bit. Um, but for where it's equal to one mod four . . .
32	*Tim:*	Right.

The 'well' that initiates Jonathan's response (29) seems to be a hedge on the maxim of Quality. After coughs and pauses, the best he can claim is 'sort of'. I soon realize the futility of this 'beating about the bush' and invite Jonathan (30) to tell me his argument, but still redressing face (his) by giving the option ('or') so that consent is not the only preferred response. He accepts the alternative, realizes that this might disappoint me and challenge me in my role as supervisor. So although his answer (31) to my question (30) is (for the moment) 'no', he presents it as 'yes', appropriately marked by a hedge on Quality ('Well'). Thus he asserts his right to present his argument in the way he chooses, but bears my prompt (28) in mind and eventually responds to it (39).

39	*Jonathan:*	Um, and then, if p is congruent to one [pause] we've then got . . . p minus, p minus one is a quadratic residue of that.
40	*Tim:*	Indeed; or minus one.
41	*Jonathan:*	Whichever way. So that gives us, um, one squared and p minus one squared, so we've got another pair of solutions.
42	*Tim:*	Umm . . . you don't quite mean that, do you? I mean, you mean, what you've got, one squared and whatever, gives you p minus one . . .
43	*Jonathan:*	Oh, yes, so . . .
44	*Tim:*	. . . when it's squared. Right, /OK/

Jonathan's account here is presented throughout in first person plural form rather than the more usual colloquial second person 'you'. The 'we's and 'us's run through his explanation (NT4:33–41). This is, of course, the classical means of rhetorical distancing, typically used by writers and speakers in formal mathematical discourse (Pimm, 1987, p. 67). Jonathan is a sufficiently sophisticated (or encultured) mathematician to deploy it. At the same time, he may be wanting to include me, to take me along with his account. I recall that this was typical of his oral contributions in lectures, if I posed a question to the class. As one of the most able mathematicians in his year group, his accounts would be in terms of 'we', I sensed, in order to include the rest of the class in his insight.

Relative to other undergraduate mathematics students, I would have said that Jonathan was quite articulate when talking about mathematics. Yet in NT4:41 he fails to say precisely what he intends to convey, and I am left to fill the gaps in the elided statement (39) ('mod 4' should follow the first 'one') and to appreciate that the demonstrative pronoun 'that' is co-referential with 'p'. The ambiguity in NT4:41 is sufficiently serious for me to feel that I have to clarify the meaning by correcting him. The consequent FTA (42) is redressed with the usual hesitation, hedging and indirectness ('you don't quite mean that, do you?').

Among the linguistic (as opposed to affective) barriers that stand between students' ideas and their articulation, two seem to be paramount. The technical language of the mathematics register seems at first to belong to the teacher; students must be enabled to inherit and appropriate it for themselves, as a mark and a measure of their enculturation. The second barrier, especially in discussion, is the syntax of 'proper' mathematical sentences, which rarely matches the cognitive production of elements of the sentence. Thus in NT4:41, Jonathan is conscious that 1 and $p - 1$ are quadratic residues ('squares') modulo p when $p \equiv 1$ mod 4. Whereas 1 is equal to the square of itself and therefore serves for 'x', $p - 1$ is not. The degree of detachment necessary, to hold the fragile mathematical idea in mind whilst the sentence is assembled with the correct syntax, is considerable.

Soon it becomes apparent that Jonathan's argument, with which he is only 'fairly happy' (9), is incomplete.

47 *Jonathan:* And, then there's this pairing thing . . .
48 *Tim:* Yeah?
49 *Jonathan:* Which . . . that's the bit I can't, I'm not . . . able to explain. I can't, I'm not, I can't say why they pair off, like that. Um, but then(?) we've got, um, p minus one over two pairs [number of quadratic residues mod p] [inaudible]
50 *Tim:* Oh, p minus one over two squares.
51 *Jonathan:* Yes. And so, so you get [long pause] Yes, sorry, yes that's it. And they add up to give p each time, these two . . . these pairs of squares . . .
52 *Tim:* Yes.
53 *Jonathan:* So you've got p there, nought.
54 *Tim:* [pause] Um, [hesitant] that's an absolutely fine . . . um, I mean, let's think, we're talking about when p is congruent with one mod four here, aren't we?

Jonathan identifies the gap in his argument, which is to show that, when $p = 1$ mod 4, the quadratic residues can always be paired to give sum zero. In NT4:54 I 'formulate' the conversation, in the sense of Garfinkel and Sacks (1970) – 'we're talking about . . .' – and try to insist in the formulation that these are his ideas ('you're saying'), not mine:

58 *Tim:* So you're saying . . . um, um . . . I'm trying to think of something that isn't [equal to 13] . . . Well, no, let's have something fairly straightforward. If you do two squared you get four, yeah?
59 *Jonathan:* Right.
60 *Tim:* And you're saying that in fact, um, thirteen minus four, or something congruent to that, minus four, mod thirteen, is always – in fact it's nine . . .
61 *Jonathan:* Yes.
62 *Tim:* . . . three squared – is always there. So you're saying, in that case, there always happen to be pairs that add to . . .
63 *Jonathan:* Yes.

I then acknowledge the gap in the argument, redressing the FTA with a positively hedged compliment 'you're absolutely right' (64, cf. 54). Jonathan's response indicates, for the first time, his discomfort. The gap in the proof is now exposed, and he can offer no resolution of the problem.

64 *Tim:* OK. I mean, can you take it any further than there? I mean, you're absolutely right. How can you take it any further than 'there always happens to be'?
65 *Jonathan:* No, I'm stumbling on this, but this is, this is the the the bit that . . . It's sort of an assumption I have to make [exhaled laugh] to go through this, and I . . .

66	*Tim:*	OK.
67	*Jonathan:*	. . . I can't . . .
68	*Tim:*	OK.
69	*Jonathan:*	I can't . . . and I know . . . or, I don't know . . . from looking at the ones that are congruent to three, mod four . . .
70	*Tim:*	Yes.
71	*Jonathan:*	. . . that there's not a constant adding up, for the pairs, so I can see that the two . . .
72	*Tim:*	Right.
73	*Jonathan:*	. . . they really do separate, but I can't explain, why they separate.

The exhaled laugh (65), like the cough, indicates Jonathan's unease. He must make an assumption to 'go through this'. 'This' is deictic, its referent an argument, hopefully not an ordeal. I adopt the minimal response strategy (66–72) (see case 3: Judith) in the hope that he may be holding something back. But in this case there is no insightful outpouring from Jonathan. Rather, he insists that he 'can't explain' (73).

My next turn (74) somewhat apologetically initiates Episode 2 and my explanation.

74	*Tim:*	OK. Well, I'd like to take you a bit further down that road, because I think you'll be quite pleased when you see it. OK?
75	*Jonathan:*	Right.
76	*Tim:*	I'm just wondering whether to talk about thirteen, or something that's less obvious. You know, 'cos . . . [Laughs]
77	*Jonathan:*	Oh dear [laughs], is thirteen obvious! [Laughs]
78	*Tim:*	No, no, no, I mean, um, an argument can be more forceful when you can't just – other than the numerical calculations – say, 'Well, obviously.' Yes?
79	*Jonathan:*	Yes.

In NT4:76, 78 I formulate the account which will follow. Since Jonathan was preparing to be a teacher, I was explicit about the didactic strategy that I was about to choose, i.e. the use of a generic example (78).

We are less than a third of the way through the transcript. I find, looking back at the text, that I was at pains to affirm Jonathan's achievements – which are significant, especially towards the end of the transcript. At the same time I have to correct errors and suggest approaches to proofs – not least because this was Jonathan's last supervision with me, and he was about to write up the project report over the Easter vacation. The elements of a polite contest can be seen in Episode 3 (reproduced below, 136–156) – compliments given (144, 150, 152) and accepted (153), Jonathan maintaining his approach, refusing sometimes to take the route I am suggesting (137), asserting his achievements (145, 147).

136	*Tim:*	OK, OK. [Pause] So which then brings us, I guess, back to x squared plus y squared equals one, does it?

137	*Jonathan:*	Uh, yes. Yes. That's a very intriguing way of . . . well, actually no, I went on to do something else first.
138	*Tim:*	Ah?
139	*Jonathan:*	Of x squared plus y squared equals n mod p, but n not being zero and having a particular value.
140	*Tim:*	OK.
141	*Jonathan:*	Um, having found out how many solutions there were for that . . .
142	*Tim:*	'Cos originally you took it to be anything other than zero . . .
143	*Jonathan:*	Yes.
144	*Tim:*	Which actually is not without interest, if I may say so.
145	*Jonathan:*	No, I think I did go on to show, um . . .
146	*Tim:*	I mean, now you know how many solutions there are for zero, you can say precisely how many there are for not-zero. [Laughs]
147	*Jonathan:*	Yes, that's basically what I'd done! [Laughs]
148	*Tim:*	OK.
149	*Jonathan:*	p squared and taken away how many other bits there are . . .
150	*Tim:*	Excellent.
151	*Jonathan:*	. . . so I've done that.
152	*Tim:*	Yes, that wraps up quite nicely too.
153	*Jonathan:*	Yes, that was very satisfying, actually.
154	*Tim:*	Yes. I think it's just a nice coincidence that you've two p minus one solutions, and when you subtract that from p squared, you get a perfect square . . .
155	*Jonathan:*	Oh, yes . . .
156	*Tim:*	p squared minus two p plus one is an algebraic square.

The encounter certainly had its lighter moments, such as when Jonathan pays me a compliment (204) in that he recognizes my own mathematical sense of curiosity. I gratefully accept (205), and Jonathan immediately eases my embarrassment (young students hardly ever praise their lecturers; doubtless they lack the confidence to do so) by teasing me (206) for my untidiness:

203	*Tim:*	Can I tell you that I don't know the answer to this? [Both laugh], I mean, simply because I've not allowed myself to think about it . . .
204	*Jonathan:*	Yes, very restrained of you. [Both laugh]. As soon as I'm out of that door you'll be going . . . [inaudible beneath Tim's laughter]
205	*Tim:*	Nice of you to suggest it. Um . . .
206	*Jonathan:*	Anything to put off sorting through that pile of papers! [Gestures at Tim's desk]
207	*Tim:*	[laughs] Dead right!

In the fifth and final episode of the conversation Jonathan surprised and delighted me with a proof which involves a neat combinatorial argument. I begin by suggesting (209) that he will find the proof too difficult. Jonathan's reply is dispreferred (210) and the FTA is marked and redressed by the three false starts. I realize that I have underrated him (211).

209	*Tim:*	You know the other thing that, um, you haven't proved – but in a way I don't feel too desperate about it because there's quite a lot around here for you to write up – is... is why there are... one more or one less than p solutions to x squared plus y squared equals one, in every case.
210	*Jonathan:*	Ah well, ah, that's, I'm coming on to that bit...
211	*Tim:*	Ah, right! Sorry...
212	*Jonathan:*	I had to backtrack to get to that.
213	*Tim:*	Oh, right. OK, OK.

His proof (NT4:215–234) is perfectly sound. Jonathan sketches it sufficiently for me to know how it is structured, and that it achieves the desired conclusion.[7] It rests in part on one of Jonathan's earlier theorems, the one that I had rather loftily consented (144) to be 'not without interest'.

Nonetheless Jonathan is quite diffident about his proof, and Shields himself (216) with some language reminiscent of the visiting scholar (case 6) – 'arm waving', 'sort of proof' cf. 'hand waving', 'almost proof':

216	*Jonathan:*	Um... I can then get back to some serious arm waving here and... and go back to my sort of proof of why there are x ... there are p plus one or p minus one solutions.

This is *his* proof, and its generality here is marked with 'you' (218, 220), whereas he uses the expository 'we' (224, 226) to refer back to results agreed earlier in the conversation.

218	*Jonathan:*	And, basically, um, you say how many... you take your mod, number...
219	*Tim:*	Right...
220	*Jonathan:*	And you work out how many possible pairs you can come up with...
221	*Tim:*	[hushed] Right...
222	*Jonathan:*	And... whatever that was, p squared...
223	*Tim:*	OK.
224	*Jonathan:*	We already know how many... solutions there are – for p congruent to one or congruent to three mod four – how many solutions there are for... x squared plus y squared is congruent to zero... [Pause]
225	*Tim:*	[hushed] Yeah...
226	*Jonathan:*	So we can get rid of those, for starters. And then we know that all the solutions that are left are divided up evenly between each of the other numbers...

My response was little short of ecstatic:

227	*Tim:*	Ohh, that's *very* nice. [Jonathan laughs] Oh, well done!
233	*Tim:*	Oh, well done, well done. Yes... [Laughs]

| 234 | *Jonathan:* | I didn't know if that was the way you were thinking of... |
| 235 | *Tim:* | That's very nice indeed. |

I urge him to write it up without delay; Jonathan responds that he has done so already.

| 241 | *Tim:* | Right. Well, I would suggest you rush away and write all this down. |
| 242 | *Jonathan:* | [laughs] Well, I wrote it all down yesterday, that ... that particular bit ... |

Both turns (241, 242) are affronts to 'face' (my imposition, Jonathan's refusal), and both are redressed in the same way, with 'Well'. There is some new work here, he acknowledges:

| 244 | *Jonathan:* | ... so it's just the, um, it's the pairing up of the quadratics when it's congruent and ... that tidying up of the, er, x squared. |

And so the supervision concluded:

247	*Tim:*	Rush away and write it up.
248	*Jonathan:*	Yes, before I forget it!
249	*Tim:*	[laughs] OK. Well that was a good way to start the week.

Jonathan did indeed write it up, and submitted his report, neatly word-processed, on time at the beginning of the Easter (summer) term. My co-examiner in the Faculty of Mathematics, on reading the script, remarked how much Jonathan seemed to have enjoyed this project; I happily agreed to his proposal that this component of the assessed work was first-class.

The written assessment was to remain incomplete. One month into that Easter term, Jonathan took his own life, alone in his house in Cambridge. He was a complex person, one who was more comfortable giving than receiving. Yet his innermost feelings he reserved for himself. I imagine that, for Jonathan, life posed a number of threats and challenges, and that, in the end, he did not believe that he could face all of them. Jonathan's college now awards an annual prize in his memory, for enthusiastic and insightful project work in mathematics.

Summary

In this chapter I have examined a number of mathematical teaching and learning situations. These demonstrate many ways in which indirect and vague language are used to support interaction in the mathematics classroom, and serve the interactional and transactional intentions of teachers and students. The same linguistic phenomena that I had identified in my own research-oriented, contingent interviews are present in these pedagogic encounters. Their presence serves to assist the definition and interpretation of the propositional attitude of speakers (students and teachers) and the dynamics of respect and politeness in a conjecturing atmosphere. In particular, I have identified in these eight episodes:

- the deictic function of the pronouns 'it' and 'you' for generalization;
- the subtle use of indirect speech acts for the redress of threats to 'face';
- the occurrence of hedges and modal forms, mainly implying uncertainty;
- reification of the Zone of Conjectural Neutrality in a conjecturing atmosphere.

More generally, I have endeavoured to demonstrate the possibility of insight into mathematical interaction by means of pragmatic analysis of discourse. In the interests of careful interpretation – so that it may be something more than 'a matter of guesswork' – I have applied the full range of the pragmatic analytical methods reviewed in Chapter 4. Classroom talk about mathematics is both transactional and interactional; interpretation of the interactional component of such talk requires attention to a wide range of human sensibilities and pragmatic goals.

Notes

1 Advanced Supplementary (AS) courses have about half the content of an Advanced level course, with the same academic rigour. The majority of students following the course at Rachel's college are very able, and studying three full A-level courses in addition to the AS, but a few may choose AS mathematics because they doubt their ability to do well on a full A-level mathematics course.

2 Within his novella *Poor Koko* (Jonathan Cape, 1974), John Fowles considers a burglar's frequent use of 'right' as a statement tag (as in the exchange 'I have very little with me' – 'Then you won't miss it, right?'). Fowles comments: 'It ["right"] may grammatically be more often an ellipsis for "Is that right?" than for "Am I right?" – but I am convinced that the psychological significance is always of the latter kind. It means in effect, I am not at all sure that I am right . . . the thing it cannot mean is self-certainty'.

3 The Ginn mathematics scheme is one of a number of such textbook schemes used in British primary schools.

4 Distance-learning materials produced by this British university for its vast number of students exploit a range of media, including videotapes as well as texts. Course EM236 is a unit of undergraduate study in Mathematics Education.

5 This is an example of stylistic transformation of the basic grammatical structures of sentences, in this case 'fronting a subordinate clause object' (Leech et al., 1982, p. 128).

6 The sequence of 'crosses' considered (at Debbie's suggestion) is interesting – not so much consecutive as recursive $(2, f(2) = 9, ff(2) = 37, fff(2) = 149)$.

7 Jonathan's proof proceeds as follows. There are p^2 ordered pairs of elements mod p; the cases $p = 4k + 1$ and $p = 4k + 3$ are then dealt with separately. In the first case there are (by Jonathan's first theorem) (NT4:156) $(p - 1)^2$ pairs with $x^2 + y^2$ not equal to zero. Moreover, the number of solutions of $x^2 + y^2 = n$ is the same for each n (NT4:157, proved in Episode 4), and n takes $p - 1$ distinct non-zero values. Hence the number of solutions for each non-zero n must be $p - 1$. The case when $p = 4k + 3$ similarly concludes with dividing $p^2 - 1$ by $p - 1$ to show that there are $p + 1$ solutions for each n.

9   Summary and Review

So what did we learn? (Margaret Thatcher, 1983)

Wisdom is the principal thing; therefore get wisdom: and with all thy getting get understanding. (Proverbs 4:7)

In this book, I have explored the proposition that, when people talk about mathematics, they use language as a means of satisfying a number of communicative 'wants'. Many of my data are drawn from expert–novice conversations, and my conclusions apply, in the first instance at least, to such discourse contexts. Broadly speaking, communicative wants in mathematics talk are of two kinds.

The first kind is co-operative and cognitive, stemming from a desire to share mathematical ideas – to give and receive insights, knowledge and understanding. It is associated with transactional functions of language. This giving and receiving is not one-way information traffic, flowing from teacher to pupil. The teacher needs to know what the pupil knows, what kind of knowledge s/he has constructed. My transcripts amply demonstrate the willingness of pupils to supply such information, and on occasion the pleasure they derive from doing so.

The second kind is of a social character, to do with establishing and sustaining relationships. It is associated with interactional functions of language. In the case of mathematics talk, language must frequently serve the cause of respect for and defence of 'face'. This is to be expected as a consequence of the asymmetry of the power relation in the expert–novice conversation, since the novice anticipates that the expert will evaluate her/his assertions. As I have observed in Chapter 6, the line between truth and error is perceived to be particularly sharp in mathematics in comparison with other subjects at school and college. But this is not a 'problem' only for novices. Experts, too, recognize the boundaries – albeit fuzzy boundaries – of their expertise, and use language to convey uncertainty to their audience when they make judgements and predictions on matters on or beyond the boundary.

In previous chapters, I have drawn attention to the following topics and issues:

- the use of pronouns by novice speakers of mathematics;
- the function of hedges in the communication of propositional attitude;
- the construct which I have called the Zone of Conjectural Neutrality;
- the development of the language of modality over the years 4 to 11;
- the use of vague and indirect language by teachers in interaction with pupils;
- pragmatic interpretation of transcripts of mathematics talk.

I shall address some of these matters just once more, and consider the implications for classroom application.

Pronouns

In the study of reference vagueness in the use of pronouns in mathematics talk (Chapter 5), the pupils' deictic intention is principally cognitive – pointing to concepts and generalizations – rather than affective. The significance of the teacher's 'we' has been considered by Mühlhäusler and Harré (1990), who propose that its function is commonly manipulative (spurious solidarity); also by Pimm (1987), who argues that its function is authoritarian (appeal to unnamed expert support). I have shown that the pupil's 'it' and 'you' are equally vague in terms of intended referent, but are not related to the teacher–pupil power imbalance. I have demonstrated association between their use and reference to concepts and generalizations. Given the importance of these referents for mathematics teaching, the significance of this connection can hardly be overstated.

The use of 'it' as a conceptual deictic enables the pupil to say what s/he could not say otherwise, to draw attention to mathematical entities whose name s/he does not know. In terms of the 'two language levels' analysis presented in Chapter 4, this pronoun (typically as object of the verb 'to do') is added to the object language as a vague variable. The notion of 'focus' as locus of attention is important here, given the claim of Moxey and Sanford (1993, p. 58) that 'pronouns are good for referring to things in focus [. . .] ease or acceptability of pronominal reference can be used as an index and a probe for the state of focus'. In the first instance, the teacher who is sensitive to the pronoun/focus connection can be made aware of the presence of a cognitive focus such as a generalization – recall Susie's 'times can do it, can't it, and add, and take . . . no, take-aways can't do it'. This teacher awareness opens up the possibility of further investigation of that focus through appropriate, contingent questioning.

This questioning could be of two kinds, which might be labelled 'conspiracy' and 'confrontation'. The conspiratorial approach is for the teacher to take up and use 'it' in the discourse as though her/his 'it' were intended to be co-referential with the pupil's. For example, 'Why can't take-aways do it, then?' would enable the confirmation or formation of hypotheses about the referent on the basis of further information about 'it' – rather as in a game of 'Twenty Questions', the only question not permitted is the name of the mystery object. Confrontation, on the other hand, amounts to an 'on record' request for the object to be revealed. For example, 'Wait a minute, what is this "it" you're talking about?' Clearly the choice of the conspiratorial or the confrontational approach must depend on a range of contextual and inter-personal factors, and there is scope for some research here.

In Chapter 5, I discussed how 'you' serves as a pointer to a general procedure or relationship. The subtle shift from 'I' to 'you' to mark a tendency towards speaker detachment is an important cognitive indicator. Oscillation between first and second person pronouns indicates a switch between action and knowledge, possibly with regard to different processes or generalizations.

The *I–you* contrast can be related to the process–object distinction (Sfard, 1991); the detachment associated with the conception of a mathematical notion as an object independent of the action of the speaker is marked by 'you'.[1] The pedagogic significance of the pupil/student's use of 'I' or 'you' may lie in the recognition of a pre- or post-objective cognitive state with regard to the mathematical notion being discussed. There is far greater ambiguity in the teacher's use of 'you' in that s/he may either be addressing the pupil (meta-language) or referring to a mathematical notion (object language). In practice the referent is pragmatically determined, and/or determined by tense and mood. Compare 'How do you find the area?' with 'How did you find the area?' and 'How would you find the area?'. My data suggest that pupils hardly ever use 'you' to address the teacher in mathematical conversations.

Modality, Hedges and Indirect Speech Acts

Hedges encode vagueness in mathematics talk, related to imprecision or uncertainty of speaker commitment. Whereas Shields clearly belong to the meta-language of mathematics talk, Approximators affect to supplement the object language and confuse its truth-conditional semantics: this perception is the motivation of G. Lakoff's classic 1973 paper on the logic of fuzzy concepts. In the end, a pragmatic (as opposed to a truth-conditional) analysis of hedges is more fruitful for the purposes of mathematics education.

Having identified hedges and epistemic modal forms as a feature of mathematics talk in a conjecturing atmosphere, my discussion of their use in Chapter 6 focused on the pragmatic goals achieved by speakers, as identified by Channell. These purposes include covering for lack of specific information and expressing politeness. In these pragmatic terms, the following important differences between pupil and 'teacher' were identified.

Essentially, pupils use epistemic hedges to shield themselves from accusation of error; the most subtle form of this is the epistemic use of Approximators, so as to render a statement 'almost unfalsifiable' by trivializing its semantics (Sadock, 1977, p. 437). The development of this aspect of communicative competence (hedging), vital to the protection of 'face' and the communication of propositional attitude in the mathematics classroom, was traced and interpreted in Chapter 7.

Teachers, on the other hand, may use hedges as a tool of pedagogic strategy, weakening the force of an assertion – their own or that of a pupil – in order to sustain pupils' engagement, emphasizing their responsibility for the determination of validity. At the same time, teachers use epistemic hedges and modal forms to perform indirect speech acts in interaction with pupils, particularly in order to present a request for information as a question – the default illocutionary force of the interrogative form. For example, 'What do you think this shape is called?' or 'Can you tell me the mean of these four numbers?' The purpose of these indirect speech acts is to alleviate the illocutionary force of an act that could lead to 'loss of face' if the pupil cannot supply the 'answer'.

The application of these findings to the teaching of mathematics is easier to state than to put into practice. The premise that knowledge is actively constructed from reflection on experience is at the heart of a constructivist view of learning. Such a view puts an onus on the teacher to try to understand the form, content and robustness of that knowledge, as an observer of and participant in pupils' mathematical activity – an 'accultured' participant, moreover, who 'can legitimise certain aspects of their mathematical activity and sanction others' (Cobb et al., 1992, p. 102). For this reason, among others, it is desirable for pupils regularly to articulate their constructed beliefs, or construct them through articulation, in the hearing of their teacher. Such a self-constructed belief may be fragile; in particular, any inductive conjecture would be expected to be. The burden of the affective baggage associated with mathematics in school then necessitates that the pupil should articulate the belief whilst distancing her/himself from full commitment to it. That is to say, pupils must convey their propositional attitude to the substance of their assertion. The rich variety, in some cases the subtlety, of hedges and modal forms deployed by pupils for this purpose is evidence of this affect-oriented dimension of pupils' communicative competence. These Markers are linguistic pointers to uncertainty and attendant cognitive vulnerability. The teacher's subtle task at such moments is to facilitate the depersonalization of the assertion, as a preliminary to 'legitimizing' or 'sanctioning', by ensuring that it is located in the Zone of Conjectural Neutrality. This will be discussed further in the next section.

My comment about the relative ease of stating such a policy as opposed to implementing it is based, in part, on my own experience. As Piaget said with reference to clinical interviewing, sensitivity to epistemic Markers in mathematics talk also 'can be learned only by long practice'. Despite my awareness of the form of such Markers, the difficulty (for me, when teaching) is in *attending* to the linguistic details of a 'lesson' whilst being fully engaged with the mathematics. I was sharply conscious of this, for example, during the delivery of the public lecture (Chapter 8). My purpose in attending had been to note linguistic features of the speaker's utterances, yet I frequently found myself diverted by growing interest in the mathematical content! This difficulty is clearly related to one of Hewitt's pedagogic aphorisms: 'The amount of attention available to us is finite and limited' (1994, p. 69). In particular, it is difficult, if not impossible, without 'long practice', to attend to two *demanding* things simultaneously. If the level of demand of one task can be reduced by automation as a result of long practice, thereby reducing its attention requirement, then simultaneous *performance* may be possible (e.g. attending to a programme on the car radio whilst driving on an open road). Perhaps some aspects of mathematics teaching can be automated in this way, but I don't know; for some time my preference has been for contingent lessons in order to *avoid* the tedium of automation.

It would, I imagine, be equally tedious to be continually and consciously preoccupied with certain syntactic features of speech in the classroom ('there goes a subordinate clause!'), even those thought to have particular pragmatic significance.

Nevertheless, I am aware of having made some personal progress in achieving the automation of sensitivity to Marked language.

For example, in a lecture to first-year university students in a course on mathematical processes, I 'invited' them to evaluate $a^2 + b$ and $b^2 + a$ when $a + b = 1$. An inductively based conjecture was proposed and proved by the students. Various 'what if?' variants followed, e.g. if $a + b = 10$, is there another 'interesting' function of a, b, symmetrical on the given subset? What if a, b are complex numbers and $a + b = 1$? The algebraic requirements of the proof indicated that we might go further. What if a and b are 2×2 matrices? After a few moments' pause one student said: 'Would they have to add up to the identity or something?' In that moment I recognized the question as an indirect speech act, asserting a tentative belief, further shielded by the vague completer 'or something' working as an epistemic Adaptor. I was very aware of the significance of what she had said, yet my attention had been on the mathematics until that moment. It was the first time I suspected that attention to a restricted range of lexical pointers (pronouns and hedges, for example) could be automated 'by long practice'. Teachers can be attuned to such pointers to propositional attitude, but more work is needed to refine their responses to pupils who articulate them in the mathematics classroom.

The Zone of Conjectural Neutrality

In introducing the construct of the ZCN in Chapter 6, I described it as a space between what we believe and what we are willing to assert. This metaphysical construct deserves a name only if it signifies something of didactic significance.

Now, the issue central to the notion of ZCN is summarized in the questions 'Where are pupils' conjectures located? Who is responsible for them?' The default position must be that a conjecture belongs to the one who utters it. If the conjecture is asserted with conviction – better still, if it is subsequently validated as true – then this is not an affective problem. But if a conjecture is offered tentatively, it is better for it to be located somewhere neutral before it is tested, in order that there may be some real prospect of Dawson's promise (1991, p. 197) of 'testing on a cognitive rather than an affective level'. I emphasize again that this is in defiance of the cultural norm that the pupil is judged to be 'right' or 'wrong' rather than the 'answer' 'true' or 'false'; that it is s/he who is on trial, not her/his beliefs.

This is at the heart of pupils' communicative competence in the use of Marked language in the assertion of conjectures. The forms of linguistic shielding which I have discussed have the effect of reifying the ZCN and locating the conjecture in it, thus distancing the speaker from the assertion that he or she makes. A Plausibility Shield such as 'I think', 'maybe', 'perhaps' does this in a very direct way, because the marker of propositional attitude lies outside the statement that follows it. Epistemic Approximators (such as 'about' in 'there are about fifteen ways') are more subtle: they do not require the speaker to disown her/his conjecture, but they do make it almost unfalsifiable. Whilst subtle, this is less than helpful, since a consequence of its vagueness is that, strictly speaking, it can be neither validated nor modified. The conventional force, however, is clearly to present the conjecture as fallible, possibly in need of modification.

The teacher who recognizes the epistemic force of a Marked conjecture has the option of assisting its placement in the ZCN. One way to do so is to write it on a flip chart or chalkboard, and say something like 'OK, let's take a look at this conjecture,' possibly without reference to the one who proposed it or constant application to him/her for arbitration or interpretation. Another way is to form small discussion groups which then tend to assume some collective ownership for the conjecture and their findings about it when reporting back to the class. I sometimes 'return' a conjecture, or an agreed modification of it, from the ZCN back to its originator when the 'severe testing' is over. I do this, for example, by marking its changed status with reference to the conjecture as 'theorem' (sometimes 'lemma') and naming it Yuko's Theorem or Tom's Theorem. If Fermat and Langrange merit such attribution,[2] then why not Yuko and Tom?

The term 'conjecture' may suggest the cognitive outcome of an extended investigation, but it could be simply the answer to a teacher's question. 'Is 91 a prime number?' 'How many non-isomorphic groups are there of order 8?' By default, the one who answers the question 'owns' the answer and is subsequently right or wrong. One way of trying to bring the answer into the ZCN before it is spoken is to pose the question as an indirect speech act. 'Can you tell me if 91 is a prime number?' As I have observed (Chapter 4) the illocutionary force of the indirect act is achieved by the modal auxiliary, which questions one of the felicity conditions (ability to comply) of the direct request for information. Another, rather different technique, is to pose questions as statements (with the tacit or explicit 'Discuss'). Thus 'Ninety-one is not a prime number'. Or by attribution: 'My friend says that 91 is a prime number.' The conjecture then goes straight into the ZCN; at the very worst, only the teacher (or his friend) is 'wrong' if the statement turns out to be false. But this technique has limitations, and cannot help with extended enquiries 'in which a conjecture is created, tested and proved, or refuted and modified' (Dawson, 1991, p. 197).

In a conjecturing atmosphere, a pupil may articulate a conjecture without necessarily being committed to its truth. Both the pupil and the teacher may influence the relocation of the conjecture from the pupil to the ZCN. The conjecture is then tested, modified or rejected in the ZCN. In such a cognitive and affective *milieu*, it is the proposition that is on trial, not the person. The ultimate goal, for the fallibilistically committed teacher, would be for the class to understand that this is the case.

Validation and Classroom Application

At our first three meetings, the eight regular members of the Informal Research Group were exposed to the linguistic framework and most of the findings reported in Chapters 5 and 6, set in the context of contingent questioning and the use of transcripts to interpret classroom interactions. The next two meetings were given over to feedback and study of the transcripts that six members of the group had made of their own tape recordings. The discussion included general consideration of the benefits of attention to vague aspects of mathematics talk in the classroom. Some of these teachers responded to my request for written reflections.

Five of the transcripts considered in Chapter 7 were donated by members of the IRG. In each case, I sent the donor a copy of my analysis of their transcript – the relevant section of Chapter 7 – inviting them to correct and/or comment on it, and asking their permission to include it. Apart from one or two minor corrections they were content that my interpretation of their transcript concurred with their recollection of their interaction with the pupil(s). Hazel's comment was typical:

> In reference to the draft chapter that you sent me, I agree with all that was written and am happy for all of it to be included. I particularly agree with two points that you made on the second page . . .

I also took it as positive that six teachers, representing four phases of schooling from 4 to 18, gave up time to be involved in the group, and to record and transcribe interviews with children, some of them lengthy texts. Their incentive appeared to be intrinsic interest and/or a sense that this might have potential for the improvement of mathematics teaching and learning in their classrooms.

In terms of the specific focus of the group on vague language, it appeared that the phenomenon of hedges was one that they could readily identify. It was also the only one. This may relate to my earlier comment about the difficulty of attending to language in the classroom as distinct from attending to the business of teaching mathematics. Nonetheless, informal feedback suggested that these teachers had been sensitized – or believed that they had been – to the use of hedges by children as an indicator of propositional attitude, principally of uncertainty. It was not always clear what they then did with the knowledge, although Hazel wrote (personal communication) that:

> I am now more aware of the effect of using vague language in the classroom, so I can use it in a positive way. [. . .] My knowledge of hedges has helped me to spot that some statements made by children are less certain than they appear. [. . .] I can then respond to them at an appropriate level.

Sue had developed the task which was the basis of her transcribed talk (five people, how many more are needed to make ten?) into one which began with (say) eight people, then seven, six, and so on, so as to offer some potential for the child to predict and generalize. She had also begun to work on the way she presented questions (in effect, as indirect speech acts) to the 4- and 5-year-olds in her class, searching for ways that would convey her attitude that their answers were modifiable conjectures. Sue wrote (personal communication) that:

> thinking about this project has influenced the way I have spoken to the children – I have only tried a small group with [the] second exercise but when I do I feel I

shall find a few more 'I think's and 'shall I try's [. . .] When maths becomes sums which are right or wrong we stifle some children's embryonic sense of pattern in number and their enthusiasm for investigation.

Perhaps Sue came closest (not only in this quotation) to articulating an understanding of the significance and possible application of the work of the IRG with regard to the communication of propositional attitude in a conjecturing atmosphere. Not surprisingly, the focus on vague language was often obscured by other, equally important insights into teaching and learning in their classrooms which the teachers gained from recording and transcribing their interactions with pupils. Judith wrote that 'This [taping and transcribing] was an insight for me into the level of my own pupils' understanding, and into some of their idiosyncrasies of thinking.' One specific instance, mentioned in Chapter 8, was Sue's realization that her intended, restricted meaning of 'more' was being widely misunderstood by young children.

Rachel listed a number of general conclusions and questions which arose from her study of her transcripts, including the following:

- The conversation is very dependent on the *rapport* between child and interviewer.
- Does the nature of the conversation change after a number of interviews?
- Is there a danger of the interviewer becoming familiar with the child's way of thinking and then beginning to interpret for them?
- Traditionally maths has been an exception to the use of children's talk as a vehicle for learning mathematics. Maths has a symbolism which seems formidable. Perhaps this has caused some teachers to base their language work on the transmission and use of symbols and on learning the formal, spoken vocabulary.
- Students often discuss things in lessons with me and each other and then say, 'But how do I write that down?'
- Maths is a precise subject – is uncertainty valid?
- I am impatient that they should get to the correct answer.
- I wonder what the students would gain from listening to the tape?
(Williams, 1995)

A comment made by Judith was illuminating. She remarked, in group discussion, that the work on vague language had been 'really interesting' but that she did not think it was useful. She couldn't see how she could use it. Hazel and Rachel (three-year veterans of teaching) responded – with all kinds of redress of FTA – that they did feel it was useful to them as teachers, and that Judith's perception might be due to her inexperience as a teacher. They recalled that they had been preoccupied with preparation and classroom management in their first year or so of teaching, and only recently had been in a position (having automated at least some aspects of their teaching?) to think about fine-tuning their interaction with pupils. Judith listened to this suggestion and agreed that it was a possible explanation. Sue and Ann (the real veterans) wisely withheld comment.

At that time I had not analysed Judith's transcript (IRG5). What is interesting, in the light of Judith's honest perception, is the richness of her conversation with Allan in terms of mathematical process (see Chapter 8: case 3) and her skill in creating and sustaining a conjecturing atmosphere. It hardly matters, for her, that she is not (yet?) able to 'use' language in an analytical way to achieve or develop what she already does intuitively. Some time later, Judith attributed her good 'instincts' to the influence of a particular tutor on her PGCE course:

> I myself was trained to look beyond language to meaning – whether I was creating an environment where students were free to make decisions and predictions, to make mistakes and correct them . . . (Personal communication)

More systematic data collection from the Informal Research Group would have been useful. In practice, I was totally dependent on the goodwill of the members of the group, since their only incentive to participate at all was personal and professional interest. Despite the fact that I set out a 'contract' for participation in the group at the beginning (attendance, supplying a transcript, contributing to a publication), I was in no position to enforce it. Nevertheless, the IRG was valuable in its affirmation of the relevance, to their classrooms, of my research on aspects of vague language.

Interpretation of Transcripts of Mathematics Talk

Many of my data are embedded in transcripts of people talking about mathematics in pedagogic or quasi-pedagogic situations. My interpretivist position denies the possibility of 'knowing', in any pure and absolute way, what a given utterance 'means'. The interpretation of the meanings and motives of others entails incorporating the evidence of the text into one's view of the world, the actors in the interaction, and one's knowledge of their situation. It is a constructive act of meaning-making for the analyst, whose 'reading' of a particular utterance must be made to fit, to be consistent with, the way they construe the utterance in its context.

Moreover, the analysis of mathematical interaction must be more than the interpretation of individual utterances; it must account for how the discourse is sustained as a social event. The participants in the conversation must make such interpretations of mathematical and social meaning in the moment, within the conversation itself, to make an interactive contribution to it. The task of the analyst, in this fullest sense, is to give, with all possible skill and integrity, an account of the record of the conversation, transformed and preserved in the transcript. Such an account is a 'story' which need not claim to be true but can endeavour to contain some *truth* about what it may be like to talk to someone – a teacher, a child, a student, a friend – about mathematics. How it feels to ask a question, or to be asked for an answer; how it is possible to say what you know, and how it feels when you do; what it's like when you know but don't know how to say it.

Perhaps, as Leech suggests (1983, pp. 30–31), interpreting an utterance is ultimately a matter of guesswork, but that does not mean that one guess is as good as another. In this book, I have set out to demonstrate that a pragmatic approach to discourse analysis offers a way of setting about this business of interpretive guesswork in a responsible way.

Conclusion

In this chapter, I have summarized the outcomes of the research in relation to the aims that I set out at the beginning. The main findings, reported in earlier chapters and reviewed above, can now be summarized as follows:

- Novice speakers of mathematics are able to make skilful use of the pronouns 'it' and 'you' to point to mathematical concepts and generalizations, and to indicate detached generalization by a subtle shift from first person to second person pronouns. The details of the associated study are reported in Chapter 5.
- Hedges play an important part in the communication of propositional attitude, and this is of vital importance in the formation and articulation of predictions and generalizations.
- In a conjecturing atmosphere, associated with fallibilistic teaching and learning, epistemic hedges implicate uncertainty. The Zone of Conjectural Neutrality is a pedagogical concept, capturing the idea of an ideal space in which tentative conjectures may ideally be tested.
- The language of modality, and of hedges in particular, develops in a more or less consistent way over the years 4 to 11. Whilst some children develop the root sense of Approximators, for most children such hedges are deployed in the institutional (school) setting to protect against accusations of error.
- These interactional dimensions of language are present across a wide age range of pupils and students in discourse with their teachers. Teachers make skilled and frequent use of indirectness to sustain the involvement and self-esteem of their pupils.
- The linguistic framework introduced in Chapter 4 is a useful one for pragmatic interpretation of transcripts of mathematics talk.

What began as a study of mathematical thinking has become a way of looking at ways of learning and styles of teaching through the particular perspective of vague language. What stands out from this pragmatic analysis is that vagueness is not a deficiency, but an essential ingredient of communicative competence in mathematical interaction. In a conjecturing atmosphere, vague language is the means by which to say what you want to say, and as much as you wish to commit yourself to, in the context in which you speak. Those of us who teach and learn mathematics would benefit from learning how to recognize such vagueness, and knowing how to use it.

Notes

1 Sfard's process–object formulation also nicely embraces the notion of 'it' as a conceptual deictic, when she writes that 'Seeing a mathematical entity as an object means being capable of referring to it as if it was a real thing' (1991, p. 4).

2 Often they don't; a case in point is 'Wilson's Theorem', which states that p is prime if and only if p divides $(p - 1)! + 1$.

Appendix 1
Transcript Conventions

Whilst a good deal of transcribed speech is presented as data in this book, I have deliberately kept the conventions used in such transcripts to a minimum. One advantage of this decision is that the text can be read more fluently. A disadvantage is the loss of many nuances of speech, particularly of prosodic data.

S3	the coded name of the transcript to which the speech belongs – all such transcripts are listed in Appendix 2.
12	number of the speaker's turn in the transcript.
S3:12	the twelfth turn in transcript S3.
John:	the name or code-name of the speaker.
[points]	description of non-linguistic communication.
[inaudible]	indecipherable speech.
[her teacher]	transcriber's situational elucidation or comment.
[...]	transcript ellipsis of words or turns.
so ... we	'...' indicates a short (untimed) pause or hesitation.
[pause]	a longer pause or silence.
so ...	utterance interrupted and/or not completed.
/ /	encloses utterance overlapping that of next or previous speaker.
think	(italic) word or words stressed by the speaker.
think	(bold) highlighted by the transcriber for the attention of the reader.

Appendix 2
Index of Transcripts

Code	Interviewer	Subject(s)	Age	Date (yy/mm/dd)
Susie				
S1	Tim	Susie	9	910422
S2	Tim	Susie	9	910429
S3	Tim	Susie	9	910513
S4	Tim	Susie	9	910520
S5	Tim	Susie	10	920115
Simon				
Si0	Tim	Simon	11	900915
Si1	Tim	Simon	12	911226
Si2	Tim	Simon	12	911229
Si3	Tim	Simon	12	911231
Make Ten				
T1	Tim	Roksana/Anna	11/10	920612
T2	Tim	Lucy/Rachel	10/10	920617
T3	Tim	Alex/Caroline	10/11	920708
T4	Tim	Harry/Alan	10/11	920708
T5	Tim	Jubair/Shofiqur	11/10	920708
T6	Tim	Ishka/Frances	10/10	920716
T7	Tim	Anthony/Sam	11/11	920716
T8	Tim	Runa/Kerry	11/11	920716
T9	Tim	Rebecca/Runi	10/11	930316
T10	Tim	Susan/Shahnaz	11/11	939316

Code	Interviewer	Subject(s)	Age	Date (yy/mm/dd)
Whole School Age-related Study of 'Markers'				
M1–45	Mark	Various	4:7–5:9	940624
M46–114	Mark	Various	5:10–7:9	940627
M115–180	Mark	Various	7:10– 9:9	940630
M181–230	Mark	Various	9:10–11:9	940707
Donated by Members of the Informal Research Group				
IRG1A	Sue	Rebecca	4:8	9503
IRG1B	Sue	Jane	4:8	9503
IRG1C	Sue	Anna	4:3	9503
IRG1A	Sue	Jason	4:11	9503
IRG2	Ann	Charlie	10	9503
IRG3	Hazel	Faye/Donna	10/10	950317
IRG4	Kevin	Andrew/Matthew	10–11	9503
IRG5	Judith	Allan	13–14	9505
IRG6A	Rachel	Juliette/Di	18/18	950308
IRG6B	Rachel	Clare/John	18/18	950308
Undergraduate Number Theory Project Supervisions				
NT1	Tim	Lorna	21	950306
NT2	Tim	Caroline	21	950307
NT3	Tim	Nicola	21	950307
NT4	Tim	Jonathan	24	950313
NT5	Tim	Lorna	21	950313
NT6	Tim	Claire	21	950313
NT7	Tim	Katy	21	950314
NT8	Tim	Nicola	21	950314
NT9	Tim	Claire	21	950317
NT10	Tim	Caroline	21	950320

References

ACKERMANN, W. (1956) 'Begrundung einer strengen Implikation', *Journal of Symbolic Logic*, **21**, pp. 113–128.

AINLEY, J. (1988) 'Perceptions of teachers' questioning styles', in Bourbàs, A. (ed.), *Proceedings of the Twelfth International Conference for the Psychology of Mathematics Education* I, Vezprém, Hungary, Ferenc Genzwein, pp. 92–99.

AINLEY, J. (1991) 'Is there any mathematics in measurement?', in Pimm, D. and Love, E. (eds), *Teaching and Learning School Mathematics*, London, Hodder and Stoughton, pp. 69–76.

Association of Teachers of Mathematics (1967) *Notes on Mathematics in Primary Schools*, Cambridge, Cambridge University Press.

Association of Teachers of Mathematics (1980) *Language and Mathematics*, Nelson, Association of Teachers of Mathematics.

ATKINSON, J. M. and DREW, P. (1979) *Order in Court*, London, Macmillan.

ATKINSON, P. (1981) 'Inspecting classroom talk', in Adelman, C. (ed.), *Uttering, Muttering*, London, Grant McIntyre, pp. 98–114.

AUSTIN, J. L. (1962a) *Sense and Sensibilia*, Warnock, G. J. (ed.), Oxford, Oxford University Press.

AUSTIN, J. L. (1962b) *How to do Things with Words*, Oxford, Oxford University Press.

BACHELARD, G. (1938) *La Formation de l'esprit scientifique*, Paris, Vrin.

BAKER, A. (1984) *A Concise Introduction to the Theory of Numbers*, Cambridge, Cambridge University Press.

BALACHEFF, N. (1988) 'Aspects of proof in pupils' practice of school mathematics', in Pimm, D. (ed.), *Mathematics, Teachers and Children*, London, Hodder and Stoughton, pp. 216–235.

BALACHEFF, N. (1990) 'Towards a *problématique* for research on mathematics teaching', *Journal for Research in Mathematics Education*, **21**(4), pp. 258–272.

BAR-HILLEL, Y. (1971) 'Out of the pragmatic wastebasket', *Linguistic Enquiry*, **2**, pp. 401–417.

BARHAM, J. C. (1988) 'Teaching Mathematics to Deaf Children', unpublished Ph.D. thesis, University of Cambridge.

BELL, A. W., COSTELLO, J. and KÜCHEMANN, D. (1983) *A Review of Research in Mathematical Education, Part A: research on learning and teaching*, Windsor, NFER-Nelson.

BERGER, P. and LUCKMANN, T. (1967) *The Social Construction of Reality*, Harmondsworth, Penguin.

BERNAYS, P. (1935) 'Sur le platonisme dans les mathématiques', *L'Enseignement mathématique*, 34, reprinted in translation in Benacerraf, P. and Putnam, H. (eds) (1964) *Philosophy of Mathematics*, Oxford, Basil Blackwell, pp. 274–286.

BERNSTEIN, B. (1971a) 'On the classification and framing of educational knowledge', in Young, M. F. D. (ed.), *Knowledge and Control: new directions for the sociology of education*, London, Collier-Macmillan, pp. 47–69.

BERNSTEIN, B. (1971b) *Class, Codes and Control* I, London, Routledge and Kegan Paul.

BERNSTEIN, R. (1983) *Beyond Objectivism and Relativism*, Philadelphia, PA, University of Pennsylvania Press.

BILLS, L. (1996) 'Shifting Sands: students' understanding of the roles of variables in "A" level mathematics', unpublished Ph.D. thesis, Milton Keynes, Open University.

BILLS, L. and ROWLAND, T. (1999) 'Examples, generalisation and proof', in Brown, L. (ed.) *First Annual Proceedings of Meetings of the British Society for Research into Learning Mathematics*, York, QED Books, pp. 103–116.

BISHOP, A. (1985) 'The social construction of meaning – a significant development for mathematics education?', *For the Learning of Mathematics*, **5**(1), pp. 24–28.

BLACK, M. (1937) 'Vagueness: an exercise in logical analysis', *Philosophy of Science*, **4**, pp. 427–455.

BRENNER, M. (1985) 'Intensive interviewing', in Brenner, M., Brown, J. and Canter, D. (eds), *The Research Interview: uses and approaches*, London, Academic Press, pp. 147–162.

BRISSENDEN, T. (1988) *Talking about Mathematics*, Oxford, Blackwell.

BROCKWAY, D. (1981) 'Semantic constraints on relevance', in Parrett, H., Sbisa, M. and Verschueren, J. (eds), *Possibilities and Limitations of Pragmatics: Proceedings of the Conference on Pragmatics at Urbino, July 8–14, 1979*, Amsterdam, Benjamins, pp. 57–78.

BROWN, G. and YULE, G. (1983) *Discourse Analysis*, Cambridge, Cambridge University Press.

BROWN, P. and LEVINSON, S. C. (1987) *Politeness: some universals in language usage*, Cambridge, Cambridge University Press.

BROWN, R. W. and GILMAN, A. (1970) 'Pronouns of power and solidarity', in Brown, R. (ed.), *Psycholinguistics*, New York, Free Press, pp. 302–336.

BROWN, S. (1981) 'Sharon's "Kye"', *Mathematics Teaching*, **94**, pp. 11–18.

BUCHANAN-BARROW, E. and BARRETT, M. (1996) 'Primary school children's understanding of school', *British Journal of Educational Psychology*, **66**, pp. 33–46.

BURGESS, R. G. (1985) 'Introduction', in Burgess, R. G. (ed.), *Strategies of Educational Research: qualitative methods*, London, Falmer Press, pp. 1–22.

BURN, R. P. (1982) *A Pathway into Number Theory*, Cambridge, Cambridge University Press.

BURNS, L. C. (1991) *Vagueness: an investigation into natural languages and the sorites paradox*, Dordrecht, Kluwer Academic Publishers.

BURROUGHS, G. E. R. (1957) *A Study of the Vocabulary of Young Children*, University of Birmingham, Institute of Education, Educational Monographs no. 1, Edinburgh, Oliver and Boyd.

BURTON, L. (1984) *Thinking Things Through: problem solving in mathematics*, Oxford, Basil Blackwell.

BUXTON, L. (1981) *Do You Panic about Mathematics?*, London, Heinemann.

CARLSON, L. (1984) *'Well' in Dialogue Games: a discourse analysis of the interjection 'well' in idealised conversation*, Amsterdam, John Benjamins.

CHAFE, W. (1972) 'Discourse structure and human knowledge' in Carroll, J. B. and Freedle, R. O. (eds), *Language Comprehension and the Acquisition of Knowledge*, Washington DC, Winston.

CHAMPAGNE, A. B., KLOPFER, L. E. and ANDERSON, J. H. (1980) 'Factors influencing the learning of classical mechanics', *American Journal of Physics*, **48**, pp. 1074–1079.

CHANNELL, J. M. (1980) 'More on approximations', *Journal of Pragmatics*, **4**, pp. 461–476.

CHANNELL, J. M. (1985) 'Vagueness as a conversational strategy', *Nottingham Linguistic Circular*, **14**, pp. 3–24.

CHANNELL, J. M. (1990) 'Precise and vague quantities in writing in economics', in Nash, W. (ed.), *The Writing Scholar: studies in the language and conventions of academic discourse*, Beverley Hills, CA, Sage Publications, pp. 95–117.

CHANNELL, J. M. (1994) *Vague Language*, Oxford, Oxford University Press.

CHEVALLARD, Y. (1985) *La Transposition didactique*, Grenoble, La Pensée Sauvage.

CHRISTIE, A. (1977) *Curtain: Poirot's Last Case*, London, Fontana Books.

CLAYTON, J. G. (1988) 'Estimation', *Mathematics Teaching*, **125**, pp. 18–19.

CLAYTON, J. G. (1992) 'Estimation in Schools', unpublished M.Phil. thesis, London, University of London.

COATES, J. (1983) *The Semantics of the Modal Auxiliaries*, London, Croom Helm.

COBB, P., YACKEL, E. and WOOD, T. (1992) 'Interaction and learning in mathematics classroom situations', *Educational Studies in Mathematics*, **23**, pp. 99–122.

CORNU, B. (1991) 'Limits', in Tall, D. (ed.), *Advanced Mathematical Thinking*, Dordrecht, Kluwer Academic Publishers.

CRYSTAL, D. (1969) *Prosodic Systems and Intonation in English*, Cambridge, Cambridge University Press.

CRYSTAL, D. (1991) *A Dictionary of Linguistics and Phonetics*, third edition, Oxford, Basil Blackwell.

CUPITT, D. (1994) 'God and the regulative ideal', letter to *The Independent*, 11 August, p. 20.

DES (1982) *Mathematics Counts*, London, HMSO.

DES (1991) *Mathematics in the National Curriculum*, London, HMSO.

DFE (1995) *Mathematics in the National Curriculum*, London, HMSO.

DAVENPORT, H. (1992) *The Higher Arithmetic*, sixth edition, Cambridge, Cambridge University Press.

References

DAWSON, S. (1991) 'Learning mathematics does not (necessarily) mean constructing the right knowledge', in Pimm, D. and Love, E. (eds), *Teaching and Learning School Mathematics*, London, Hodder and Stoughton, pp. 195–204.

DELDERFIELD, R. F. (1969) *Come Home, Charlie, and Face Them*, London, Hodder and Stoughton.

DEWEY, J. (1923) 'The pragmatism of Peirce', a supplementary essay in Cohen, M. R. (ed.) *Chance, Love and Logic*, London, Kegan Paul, pp. 301–308.

DONALDSON, M. (1978) *Children's Minds*, London, Fontana.

DREW, P. (1981) 'Adults' corrections of children's mistakes', in French, P. and MacLure, M. (eds), *Adult–Child Conversation*, London, Croom Helm, pp. 244–268.

DREW, P. and HERITAGE, J. (1992) 'Analyzing talk at work: an introduction', in Drew, P. and Heritage, J. (eds) *Talk at Work: interaction in institutional settings*, Cambridge, Cambridge University Press, pp. 3–63.

DUBOIS, B. L. (1987) '"Something on the order of around forty to forty-four": imprecise numerical expressions in biomedical slide talks', *Language in Society*, **16**, pp. 527–541.

DUBOIS, J. (1969) *Grammaire structurale du français: la phrase et les transformations*, Paris, Larousse.

DURKIN, K. and SHIRE, B. (1991) 'Lexical ambiguity in mathematics contexts', in Durkin, K. and Shire, B. (eds), *Language in Mathematical Education*, Milton Keynes, Open University Press, pp. 71–85.

EDWARDS, R. P. A. and GIBBON, V. (1973) *Words Your Children Use*, London and Toronto, Burke Books.

EISENHART, M. A. (1988) 'The ethnographic research tradition and mathematics education research', *Journal for Research in Mathematics Education*, **19**(2), pp. 99–114.

ELLIS, B. (1968) *Basic Concepts of Measurement*, Cambridge, Cambridge University Press.

ERNEST, P. (1983) 'Thinking of a funny line', *Mathematical Education for Teaching*, **4**(2), pp. 30–40.

ERNEST, P. (1991) *The Philosophy of Mathematics Education*, London, Falmer Press.

ERNEST, P. (1998) *Social Constructivism as a Philosophy of Mathematics*, Albany, NY, SUNY Press.

FARNHAM, D. (1975) 'Language and mathematical understanding', *Recognitions*, **3**, Derby, Association of Teachers of Mathematics.

FAUVEL, J. (1987) *The Greek Concept of Proof*, Block 1, Unit 3, OU course MA290, Milton Keynes, Open University Press.

FAWCETT, R. P. and PERKINS, M. R. (1980) *Child Language Transcripts 6–12, with a preface*, Pontypridd, Polytechnic of Wales, Department of Behavioural and Communication Studies.

FIELKER, D. (1988) 'Metaphors and models', *Mathematics Teaching*, **124**, pp. 4–6.

FISCH, M. H. and TURQUETTE, A. R. (1966) 'Peirce's triadic logic', *Transactions of the Charles S. Peirce Society*, **2**, pp. 71–85.

FISCHBEIN, E. (1987) *Intuition in Science and Mathematics: an educational approach*, Dordrecht, Reidel.

FISCHBEIN, E. (1990) 'Introduction', in Nesher, P. and Kilpatrick, J. (eds), *Mathematics and Cognition: a research synthesis by the International Group for the Psychology of Mathematics Education*, Cambridge, Cambridge University Press, pp. 1–13.

FLETCHER, T. J. (ed.) (1969) *Some Lessons in Mathematics*, Cambridge, Cambridge University Press.

FOXMAN, D. D., CRESSWELL, M. J. and BADGER, M. E. (1981) *Mathematical Development: primary survey report no. 2*, London, HMSO.

FOXMAN, D. D., CRESSWELL, M. J., WARD, M., BADGER, M. E., TUSON, J. A. and BLOOMFIELD, B. A. (1980) *Mathematical Development: primary survey report no. 1*, London, HMSO.

FOXMAN, D. D., RUDDOCK, G. J., BADGER, M. E. and MARTINI, R. M. (1982) *Mathematical Development: primary survey report no. 3*, London, HMSO.

FRANKEL, R. (1990) 'Talking in interviews: a dispreference for patient-initiated questions in physician–patient interviews', in Psathas, G. (ed.) *Studies in Ethnomethodology and Conversation Analysis*, Lanham, MD, University Press of America, pp. 231–262.

FREUDENTHAL, H. (1978) *Weeding and Sowing: preface to a science of mathematical education*, Dordrecht, Reidel.

FREUDENTHAL, H. (1991) *Revisiting Mathematics Education: the China lectures*, Dordrecht, Kluwer Academic Publishers.

FUSON, K. C. (1988) *Children's Counting and Concepts of Number*, New York, Springer Verlag.

FUSON, K. C. (1991) 'Children's early counting: saying the number–word sequence, counting objects, and understanding cardinality', in Durkin, K. and Shire, B. (eds), *Language in Mathematical Education*, Milton Keynes, Open University Press, pp. 27–39.

GARFINKEL, H. and SACKS, H. (1970) 'On formal structures of practical actions', in McKinney, C. and Tiriakian, E. (eds), *Theoretical Sociology*, New York, Appleton Century Crofts, pp, 337–366.

GARROD, S. C. and SANFORD, A. J. (1982) 'The mental representation of discourse in a focused memory system', *Journal of Semantics*, **1**, pp. 21–41.

GATTEGNO, G. (1981) 'Children and mathematics: a new appraisal', *Mathematics Teaching*, **94**, pp. 5–7.

GAZDAR, G. (1979) *Pragmatics: implicature, presupposition and logical form*, New York, Academic Press.

GELMAN, R. (1977) 'How young children reason about small numbers', in Castellan, N. J., Pisoni, D. B. and Potts, G. R. (eds), *Cognitive Theory* II, Hillsdale, NJ, Lawrence Erlbaum Associates, pp. 219–238.

GELMAN, R. and GALLISTEL, C. R. (1978) *The Child's Understanding of Number*, Cambridge, MA, Harvard University Press.

GIARELLI, J. B. (1988) 'Qualitative inquiry in philosophy and education: notes on a pragmatic tradition', in Sherman, R. R. and Webb, R. B. (eds), *Qualitative Research in Education: focus and methods*, London, Falmer Press, pp. 22–27.

GINSBURG, H. P. (1977) *Children's Arithmetic: the learning process*, New York, Van Nostrand.

GINSBURG, H. P. (1981) 'The clinical interview in psychological research on mathematical thinking: aims, rationales, techniques', *For the Learning of Mathematics*, **1**(3), pp. 4–11.

GINSBURG, H. P. and RUSSELL, R. L. (1981) 'Social class and racial influences on early mathematical thinking', *Monographs of the Society for Research in Child Development*, **46**(6), pp. 32–51.

GINSBURG, H. P., KOSSAN, N. E., SCHWARTZ, R. and SWANSON, D. (1983) 'Protocol methods in research on mathematical thinking', in Ginsburg, H. P. (ed.), *The Development of Mathematical Thinking*, New York, Academic Press, pp. 7–47.

GLASER, B. G. and STRAUSS, A. L. (1967) *The Discovery of Grounded Theory: strategies for qualitative research*, Chicago, Aldine Publishing.

GLASERSFELD, E. VON (1989) 'Constructivism in education', in Husen, T. and Postlethwaite, T. N. (eds), *The International Encyclopedia of Education*, Supplementary Volume, Oxford, Pergamon Press.

GLASERSFELD, E. VON (1995) *Radical Constructivism: a way of knowing and learning*, London, Falmer Press.

GOFFMAN, E. (1967) *Interaction Ritual: essays on face to face behavior*, New York, Doubleday Anchor.

GOGUEN, J. A. (1969) 'The logic of inexact concepts', *Synthèse*, **19**, pp. 325–373.

GOODY, E. (ed.) (1978) *Questions and Politeness: strategies in social interaction*, Cambridge, Cambridge University Press.

GORDON, D. and LAKOFF, G. (1971) 'Conventional postulates', *Papers from the Seventh Regional Meeting of the Chicago Linguistics Society*, Chicago, Chicago Linguistic Society, pp. 63–84.

GRAHAM, D. (1993) *A Lesson for us all: the making of the National Curriculum*, London, Routledge.

GRANVILLE, G. and KATZ, I. (1993) 'The number's up for maths' greatest riddle', *The Guardian*, 24 June, Section 2, pp. 2–3.

GRICE, H. P. (1975) 'Logic and conversation', in Cole, P. and Morgan, J. L. (eds), *Syntax and Semantics* III, *Speech Acts*, New York, Academic Press, pp. 41–58.

GRICE, H. P. (1989) *Studies in the Way of Words*, Cambridge, MA, Harvard University Press.

GRIFFITHS, H. B. (1983) 'Simplification and complexity in mathematics education', *Educational Studies in Mathematics*, **14**(3), pp. 297–317.

HALLIDAY, M. A., McINTOSH, A. and STEVENS, P. (1964) *The Linguistic Sciences and Language Teaching*, London, Longman.

HALLIDAY, M. A. K. (1976) 'Modality and modulation in English', in Kress, G. R. (ed.) *Halliday: system and function in language*, London, Oxford University Press, pp. 189–213.

HARDCASTLE, L. and ORTON, T. (1993) 'Do they know what we are talking about?', *Mathematics in School*, **22**(3), pp. 12–14.

HARDY, G. H. (1940) *A Mathematician's Apology*, Cambridge, Cambridge University Press. (Reprinted 1967 with a foreword by C. P. Snow.)

HAREL, G. and TALL, D. (1991) 'The general, the abstract, and the generic in advanced mathematics', *For the Learning of Mathematics*, **11**(1), pp. 38–42.

HART, K. M. (1979) 'Ratio and proportion', in Hart, K. M. (ed.), *Children's Understanding of Mathematics 11–16*, London, John Murray, pp. 88–102.

HEMPEL, C. G. (1965) *Aspects of Scientific Explanation*, New York, Free Press.

HERITAGE, J. (1984) *Garfinkel and Ethnomethodology*, Cambridge, Polity Press.

HERSH, R. (1979) 'Some proposals for reviving the philosophy of mathematics', *Advances in Mathematics*, **31**, pp. 31–50.

HERSH, R. (1993) 'Proving is convincing and explaining', *Educational Studies in Mathematics*, **24**, pp. 389–399.

HEWITT, D. (1988) 'Postscript', *Mathematics Teaching*, **123**, p. 61.

HEWITT, D. (1992) 'Train spotters' paradise', *Mathematics Teaching*, **140**, pp. 6–8.

HEWITT, D. (1994) 'The Principle of Economy in the Learning and Teaching of Mathematics', unpublished Ph.D. thesis, Milton Keynes, Open University.

HEWITT, D. (1997) 'Teacher as amplifier, teacher as editor', in Pehkonen, E. (ed.) *Proceedings of the Twenty-first Conference of the International Group for the Psychology of Mathematics* III, Finland, University of Helsinki, pp. 37–80.

HEYTING, A. (1964) 'The intuitionistic foundations of mathematics', in Benacerraf, P. and Putnam, H. (eds) (1964) *Philosophy of Mathematics*, Oxford, Basil Blackwell, pp. 42–49.

HODGE, R. and KRESS, G. (1988) *Social Semiotics*, Cambridge, MA, Polity Press.

HOFLAND, K. and JOHANSSON, S. (1982) *Word Frequencies in British and American English*, Oslo, Norwegian Computing Centre for the Humanities.

HOLLAND, J. H., HOLYOAK, K. J., NISBETT, R. E. and THAGARD, P. R. (1986) *Induction: processes of inference, learning and discovery*, Cambridge, MA, MIT Press.

HOLT, J. (1969) *How Children Fail*, Harmondsworth, Penguin Books.

HOPPER, R. (1989) 'Conversation analysis and social psychology as descriptions of interpersonal communication', in Roger, D. and Bull, P. (eds) *Conversation: an interdisciplinary perspective*, Clevedon, Multilingual Matters, pp. 48–65.

HORN, L. (1984) 'Toward a new taxonomy for pragmatic inference', in Schiffrin, D. (ed.), *Georgetown Round Table on Languages and Linguistics 1984*, New York, Georgetown University Press, pp. 11–42.

HOWES, D. (1966) 'A word count of spoken English', *Journal of Verbal Learning and Verbal Behaviour*, **5**, pp. 572–604.

HYMES, D. (1972) 'Introduction', in Cazden, C. B., John, V. P. and Hymes, D. (eds) *Functions of Language in the Classroom*, New York, Teachers College Press, pp. xi–lvii.

INHELDER, B. and PIAGET, J. (1958) *The Growth of Logical Thinking from Childhood to Adolescence*, New York, Basic Books.

JENSEN, E. M., REESE, E. P. and REESE, T. W. (1950) 'The subitising and counting of visually presented fields of dots', *Journal of Psychology*, **30**, pp. 363–392.

JOFFE, L. (undated, *c.* 1985) *Practical Testing in Mathematics at Age 11*, London, Assessment of Performance Unit.

JOHNSON, P. (1970) 'A different value system and its effect on teaching methods', *Mathematics Teaching*, **53**, pp. 50–52.

KANE, R. B., BYRNE, M. A. and HATER, M. A. (1974) *Helping Children Read Mathematics*, New York, American Book Company.

KEATS, J. (1820) 'Ode to a Grecian Urn', in *Lamia, Isabella, the Eve of St Agnes and other poems*, London, Taylor and Hessey.

KILPATRICK, J. (1988) 'Editorial', *Journal for Research in Mathematics Education*, **19**(2), p. 98.

KOSKO, B. (1994) *Fuzzy Thinking: the new science of fuzzy logic*, London, HarperCollins.

LABERGE, S. and SANKOFF, G. (1980) 'Anything you can do', in Sankoff, G. (ed.), *The Social Life of Language*, Philadelphia, University of Pennsylvania Press, pp. 271–294.

LABORDE, C. (1979) 'Audacity and reason: French research in mathematics education', *For the Learning of Mathematics*, **9**(3), pp. 31–36.

LABOV, W. (1970) 'The study of language in its social context', *Studium Generale*, **23**, pp. 30–87.

LABOV, W. and FANSHEL, D. (1977) *Therapeutic Discourse: psychotherapy as conversation*, New York, Academic Press.

LAKATOS, I. (1976) *Proofs and Refutations: the logic of mathematical discovery*, Cambridge, Cambridge University Press.

LAKOFF, G. (1972) 'Hedges: a study in meaning criteria and the logic of fuzzy concepts', *Papers from the Eighth Regional Meeting of the Chicago Linguistics Society*, Chicago, Chicago Linguistic Society, pp. 183–228.

LAKOFF, G. (1973) 'Hedges: a study in meaning criteria and the logic of fuzzy concepts', *Journal of Philosophical Logic*, **2**, pp. 458–508.

LAKOFF, R. (1973) 'Questionable answers and answerable questions', in Kachru, B. B. et al. (eds), *Issues in Linguistics: papers in honor of Henry and Renée Kahane*, Urbana, IL, University of Illinois Press, pp. 453–467.

LAKOFF, R. (1975) *Language and Woman's Place*, New York, Harper.

LAPLACE, P. S. (1902) *A Philosophical Essay on Probabilities*, New York, Truscott and Emory.

LEECH, G. (1983) *Principles of Pragmatics*, London, Longman.

LEECH, G., DEUCHAR, M. and HOOGENRAAD, R. (1982) *English Grammar for Today: a new introduction*, London, Macmillan.

LERMAN, S. (1989) 'Constructivism, mathematics and mathematics education', *Educational Studies in Mathematics*, **20**, pp. 211–223.

LEVINSON, S. C. (1983) *Pragmatics*, Cambridge, Cambridge University Press.

LOVE, E. (1988) 'Evaluating mathematical activity', in Pimm, D. (ed.), *Mathematics, Teachers and Children*, London, Hodder and Stoughton, pp. 249–263.

LUKASIEWICZ, J. (1920) 'O logice trójwartosciowej' [On three-valued logic], *Ruch Filozoficzny*, **5**, pp. 169–171.

LUNZER, E. (1978) 'Formal reasoning: a reappraisal', in Presseisen, B. Goldstein, D. and Appel, M. H. (eds), *Language and Operational Thought*, II, *Topics in Cognitive Development*, New York, Plenum, pp. 47–77.

LYONS, J. (1968) *Introduction to Theoretical Linguistics*, Cambridge, Cambridge University Press.

LYONS, J. (1977) *Semantics*, Cambridge, Cambridge University Press.

MACKIE, A. (1923) 'Psycho-analysis and education', *Australasian Journal of Psychology and Philosophy*, 1, pp. 105–110.

MacLURE, M. and FRENCH, P. (1980) 'Routes to right answers: on pupils' strategies for answering teachers' questions', in Woods, P. (ed.), *Pupil Strategies: explorations in the sociology of the school*, London, Croom Helm, pp. 74–93.

MacWHINNEY, B. (1991) *The CHILDES Project: tools for analyzing talk*, Hillsdale, NJ, Lawrence Erlbaum Associates.

MAHER, C. A., MARTINO, M. M. and DAVIS, R. B. (1994) 'Children's different ways of thinking about fractions', in da Ponte, J. and Matos, J. F. (eds) *Proceedings of PME-18* III, Lisbon, University of Lisbon, pp. 208–215.

MASON, J. (1988) *Learning and Doing Mathematics*, London, Macmillan.

MASON, J. and PIMM, D. (1984) 'Generic examples: seeing the general in the particular', *Educational Studies in Mathematics*, 15, pp. 277–289.

MEY, J. (1993) *Pragmatics*, Oxford, Basil Blackwell.

MILL, J. S. (1843) *A System of Logic Ratiocinative and Inductive: being a connected view of the principles of evidence and the methods of scientific investigation*, fourth edition, London, John W. Parker.

MORGAN, C. (1998) *Writing Mathematically: the discourse of investigation*, London: Falmer Press.

MORRIS, C. W. (1938) 'Foundations of the theory of signs', in Neurath, O., Carnap, R. and Morris, C. (eds), *International Encyclopaedia of Unified Science*, Chicago, University of Chicago Press.

MOXEY, L. M. and SANFORD, A. J. (1993) *Communicating Quantities: a psychological approach*, Hillsdale, NJ, Lawrence Erlbaum Associates.

MÜHLHÄUSLER, P. and HARRÉ, R. (1990) *Pronouns and People: the linguistic construction of social and personal identity*, Oxford, Basil Blackwell.

NIVEN, I. and ZUCKERMAN, H. S. (1980) *An Introduction to the Theory of Numbers*, fourth edition, New York, John Wiley.

NELSON, R. B. (1993) *Proofs without Words: exercises in visual thinking*, Washington, DC, Mathematical Association of America.

NUNES, T. (1996) 'Language and the socialisation of thinking', in Puig, L. and Gutierrez, A., *Proceedings of the Twentieth Conference of the International Group for the Psychology of Mathematics Education* I, University of Valencia, Spain, pp. 71–76.

ORTON, A. (1985) *Studies in Mechanics Learning*, Leeds, University of Leeds Centre for Studies in Science and Mathematics Education.

OTTERBURN, M. K. and NICHOLSON, A. R. (1976) 'The language of CSE mathematics', *Mathematics in School*, 5(5), pp. 18–20.

PALEY, V. G. (1981) *Wally's Stories*, Cambridge, MA. Harvard University Press.

PEIRCE, C. S. (1902) 'Vague', in Baldwin, J. M. (ed.), *Dictionary of Philosophy and Psychology*, London, Macmillan.

PEIRCE, C. S. (1923) *Chance, Love and Logic*, ed. Cohen, M. R., London, Kegan Paul.

PEIRCE, C. S. (1932) *Collected Papers* II, *Elements of Logic*, ed. Hartshorne, C., Weiss, P. and Burks, A., Cambridge, MA, Harvard University Press.

PEIRCE, C. S. (1934) *Collected Papers* V, *Pragmatism and Pragmaticism*, ed. Hartshorne, C., Weiss, P. and Burks, A., Cambridge, MA, Harvard University Press.

PERERA, K. (1990) 'Grammatical differentiation between speech and writing in children aged 8 to 12', in Carter, R. (ed.) *Knowledge about Language*, London, Hodder and Stoughton, pp. 216–234.

PIAGET, J. (1929) *The Child's Conception of the World*, New York, Harcourt Brace.

PIAGET, J. (1937) *La Construction du réel chez l'enfant* [The construction of reality in the child, translation, M. Cook, New York, Basic Books, 1971], Neuchâtel, Delachaux et Niestlé.

PIAGET, J. (1952) 'Jean Piaget', in Boring, E. G. et al. (eds), *History of Psychology in Autobiography* IV, Worcester, MA, Clark University Press, pp. 237–256.

PIAGET, J. (1970) *Le Structuralisme*, fourth edition, Paris, Presses Universitaires de France.

PIAGET, J. and SZEMINSKA, A. (1952) *The Child's Conception of Number*, London, Routledge and Kegan Paul.

PIÉRRAULT-LE BONNIEC, G. (1980) *The Development of Modal Reasoning: genesis of necessity and possibility notions*, London, Academic Press.

PIMM, D. (1981) 'Metaphor and analogy in mathematics', *For the Learning of Mathematics*, 1(3), pp. 47–50.

PIMM, D. (1986) 'Beyond reference', *Mathematics Teaching*, 116, pp. 48–51.

PIMM, D. (1987) *Speaking Mathematically: communication in mathematics classrooms*, London, Routledge and Kegan Paul.

PIMM, D. (1992) 'Classroom language and the teaching of mathematics', in Nickson, M. and Lerman, S. (eds), *The Social Context of Mathematics Education: theory and practice*, London, South Bank University, pp. 67–81.

PIMM, D. (1994) 'Mathematics classroom language: form, function and force', in Biehler, R., Scholz, R., Straßer, R. and Winkelmann, B. (eds), *Didactics of Mathematics as a Scientific Discipline*, Dordrecht, Kluwer Academic Publishers, pp. 159–169.

PIRIE, S. E. B., MARTIN, L. and KIERAN, T. (1994) 'Mathematical images for fractions: help or hindrance?', in da Ponte, J. and Matos, J. F. (eds), *Proceedings of PME-18* III, Lisbon, University of Lisbon, pp. 247–254.

PLUNKETT, S. (1979) 'Icons', *Mathematics Teaching*, 86, pp. 6–7.

POLYA, G. (1945) *How to Solve it: a new aspect of mathematical method*, Princeton, NJ, Princeton University Press.

POLYA, G. (1954a) *Mathematics and Plausible Reasoning* I, *Induction and Analogy in Mathematics*, Princeton, NJ, Princeton University Press.

POLYA, G. (1954b) *Mathematics and Plausible Reasoning* II, *Patterns of Plausible Inference*, Princeton, NJ, Princeton University Press.

POLYA, G. (1962) *Mathematical Discovery: on understanding, learning and teaching problem solving* I, New York, John Wiley.

POLYA, G. (1965) *Mathematical Discovery: on understanding, learning and teaching problem solving* II, New York, John Wiley.

POLYA, G. (1992) quoted in *For the Learning of Mathematics*, **12**(2), p. 11.

PRINCE, E. F., FRADER, J. and BOSK, C. (1982) 'On hedging in physician–physician discourse', in di Pietro, R. J. (ed.), *Linguistics and the Professions*, Norwood, NJ, Ablex, pp. 83–96.

REES, A. (1983) 'Pronouns of Person and Power', unpublished M. A. dissertation, University of Sheffield.

RESCHER, N. (1979) *Cognitive Systematization: a systems-theoretic approach to a coherentist theory of knowledge*, Oxford, Basil Blackwell.

RESCHER, N. (1980) *Induction: an essay on the justification of inductive reasoning*, Oxford, Basil Blackwell.

RINSLAND, H. D. (1945) *A Basic Vocabulary of Elementary School Children*, New York, Macmillan.

ROLF, B. (1981) *Topics on Vagueness*, Lund, Sweden, Lunds Universitet.

RORTY, R. (1980) *Philosophy and the Mirror of Nature*, Oxford, Basil Blackwell.

ROSCH, E. (1975) 'Cognitive reference points', *Cognitive Psychology*, **104**, pp. 192–233.

ROWLAND, T. (1974) 'Real functions which generate the dihedral groups', *Mathematics Teaching*, 69, pp. 40–47.

ROWLAND, T. (1982) 'Teaching directed numbers: an experiment', *Mathematics in School*, **11**(1), pp. 24–27.

ROWLAND, T. (1990) 'Apparatus for number work', *CAN Newsletter* 2, Cambridge, CAN Continuation Project, Homerton College, pp. 1–2.

ROWLAND, T. (1992a) 'Pointing with pronouns', *For the Learning of Mathematics*, **12**(2), pp. 44–48.

ROWLAND, T. (1992b) 'Pop maths postscript', *Mathematical Gazette*, **76**(476), pp. 254–256.

ROWLAND, T. (1993) ' "Hyperpainting" – mind games for night riders', *Mathematics Teaching*, **143**, pp. 28–30.

ROWLAND, T. (1994) *CAN in Suffolk: the first six months of a calculator-aware number curriculum*, Homerton Research Reports series, Cambridge, Publication Unit, Homerton College.

ROWLAND, T. (1995a) 'Between the lines: the languages of mathematics', in Anghileri, J. (ed.), *Children's Mathematical Thinking in the Primary Years*, London, Cassell, pp. 54–73.

ROWLAND, T. (1995b) 'Hedges in mathematics talk: linguistic pointers to uncertainty', *Educational Studies in Mathematics*, **29**(4), pp. 327–353.

ROWLAND, T. (1996) 'Researching mathematical thinking with talk, tape and transcripts', in Smart, T. (ed.) *Proceedings of the Third British Congress in Mathematics Education*, Manchester, Manchester Metropolitan University, pp. 279–286.

ROWLAND, T. (1997a) 'Dividing by three-quarters: what Susie saw', *Mathematics Teaching*, **160**, pp. 30–33.

Rowland, T. (1997b) 'Fallibilism and the zone of conjectural neutrality', in Pehkonen, E. (ed.) *Proceedings of the Twenty-first Conference of the International Group for the Psychology of Mathematics* IV, Finland, University of Helsinki, pp. 80–87.

Russell, B. A. W. (1923) 'Vagueness', *Australasian Journal of Psychology and Philosophy*, **1**, pp. 84–92.

Sadock, J. M. (1977) 'Truth and approximations', *Berkeley Linguistic Society Papers*, **3**, pp. 430–439.

Saran, R. (1985) 'The use of archives and interviews in research on educational policy', in Burgess, R. G. (ed.), *Strategies of Educational Research: qualitative methods*, London, Falmer Press, pp. 207–241.

Saxe, G. B. and Kaplan, R. (1981) 'Gesture in early counting: a developmental analysis', *Perceptual and Motor Skills*, **53**, pp. 851–854.

Schegloff, E. A. (1987) 'Between macro and micro: contexts and other connections', in Alexander, J., Giesen, B., Munch, R. and Smelser, N. (eds), *The Macro-micro Link*, Berkeley, CA, University of California Press, pp. 207–234.

Schegloff, E. A. and Sacks, H. (1973) 'Opening up closings', *Semiotica* **7**, pp. 289–327.

Schiffrin, D. (1987) *Discourse Markers*, Cambridge, Cambridge University Press.

Schiffrin, D. (1994) *Approaches to Discourse*, Oxford, Basil Blackwell.

Schumann, H. and Green, D. (1994) *Discoveing Geometry with a Computer – using Cabri Géomètre*, Bromley, Chartwell-Bratt.

Searle, J. R. (1969) *Speech Acts*, Cambridge, Cambridge University Press.

Searle, J. R. (1975) 'Indirect speech acts', in Cole, P. and Morgan, J. L. (eds), *Syntax and Semantics* III, *Speech Acts*, New York, Academic Press, pp. 59–82.

Semadeni, Z. (1984) 'Action proofs in primary mathematics teaching and in teacher training', *For the Learning of Mathematics*, **4**(1), pp. 32–34.

Sewell, B. (1985) *Use of Mathematics by Adults in Daily Life*, Leicester, Advisory Council for Adult and Continuing Education.

Sfard, A. (1991) 'On the dual nature of mathematical conceptions', *Educational Studies in Mathematics*, **22**, pp. 1–36.

Shuard, H. (1986) *Primary Mathematics Today and Tomorrow*, York, Longman/SCDC.

Shuard, H. and Rothery, A. (1984) *Children Reading Mathematics*, London, John Murray.

Siegel, W. S., Goldsmith, L. T. and Madson, C. R. (1982) 'Skill in estimation problems of extent and numerosity', *Journal for Research in Mathematics Education*, **13**(3), pp. 211–232.

Simons, H. (1981) 'Conversation piece: the practice of interviewing in case study research', in Adelman, C. (ed.) *Uttering, Muttering*, London, Grant McIntyre, pp. 27–51.

Sinclair, J. McH. and Coulthard, R. M. (1975) *Towards an Analysis of Discourse: the English used by teachers and pupils*, Oxford, Oxford University Press.

Singh, S. (1997) *Fermat's Last Theorem*, London, Fourth Estate.

SKEMP, R. R. (1979) *Intelligence, Learning and Action*, Chichester, John Wiley.

SOWDER, J. (1992) 'Estimation and Number Sense', in Grouws, D. A. (ed.), *Handbook of Research on Mathematics Teaching and Learning*, New York, Macmillan, pp. 371–389.

SPERBER, D. and WILSON, D. (1986a) *Relevance: communication and cognition*, Oxford, Basil Blackwell.

SPERBER, D. and WILSON, D. (1986b) 'Loose talk', *Proceedings of the Aristotelian Society*, **86**, pp. 153–171.

STAMP, M. (1984) 'Perimeter equals area', in *Topics in Mathematics*, Derby, Association of Teachers of Mathematics, p. 3.

STEFFE, L. P. (1991) 'The constructivist teaching experiment: illustrations and implications', in von Glasersfeld, E. (ed.), *Radical Constructivism in Mathematics Education*, Dordrecht, Kluwer Academic Publishers, pp. 177–194.

STEFFE, L. P. and COBB, P. (1988) *Construction of Arithmetical Meanings and Strategies*, New York, Springer Verlag.

STEFFE, L. P., VON GLASERSFELD, E., RICHARDS, J. and COBB P. (1983) *Children's Counting Types: philosophy, theory and application*, New York, Praeger.

STEPHANY, U. (1986) 'Modality', in Fletcher, P. and Garman, M. (eds), *Language Acquisition*, second edition, Cambridge, Cambridge University Press.

STERN, G. (1964) *Meaning and Change of Meaning*, Bloomington, IN, Indiana University Press.

STUBBS, M. (1986) '"A matter of prolonged fieldwork": notes towards a modal grammar of English', *Applied Linguistics*, 7(1), pp. 1–25.

STUBBS, M. (1996) *Text and Corpus Linguistics*, Oxford, Basil Blackwell.

THOMPSON, I. (ed.) (1997) *Teaching and Learning early Number*, Milton Keynes, Open University.

TIEGEN, K. M. (1990) 'To be convincing or to be right: a question of preciseness', in Gilhooly, K. J., Keane, M. T. G., Logie, R. H. and Erdos, G. *Lines of Thinking: reflections on the psychology of thought*, Chichester, John Wiley, pp. 299–314.

TIROSH, D. (1991) 'The role of students' intuitions of infinity in teaching the Cantorial theory', in Tall, D. (ed.), *Advanced Mathematical Thinking*, Dordrecht, Kluwer Academic Publishers, pp. 199–214.

ULLMANN, S. (1962) *Semantics: an introduction to the science of meaning*, Oxford, Basil Blackwell.

VAN DEN BRINK, J. (1984) 'Acoustic counting and quantity counting', *For the Learning of Mathematics*, 4(2), pp. 2–13.

WALES, K. M. (1980) 'Exophora re-examined: the uses of the personal pronoun WE in present-day English', *UEA Papers in Linguistics*, **12**, pp. 21–44.

WALES, R. (1986) 'Deixis', in Fletcher, P. and Garman, M. (eds), *Language Acquisition*, second edition, Cambridge, Cambridge University Press, pp. 401–428.

WALKERDINE, V. (1988) *The Mastery of Reason: cognitive development and the production of rationality*, London, Routledge.

WALLWORK, J. F. (1969) *Language and Linguistics*, London, Heinemann.

WALTHER, G. (1984) 'Action proof vs illuminating examples?', *For the Learning of Mathematics*, **4**(3), pp. 10–12.

WATSON, A. (1994) 'My classroom', in Bloomfield, A. and Harries, T. (eds) *Teaching, Learning and Mathematics*, Derby, Association of Teachers of Mathematics.

WATSON, A. (1995) 'Generalising', personal communication.

WATSON, R. (1994) 'Runs of composite integers and the Chinese Remainder Theorem', *Mathematical Gazette*, **78**(482), pp. 167–172.

WEINER, B. (1972) 'Attribution theory, achievement motivation, and the educational process', *Review of Educational Research*, **42**, pp. 203–215.

WELLS, C. G. (1979) 'Learning and using the auxiliary verb in English', in Lee, V. (ed.), *Language Development*, London, Croom Helm, pp. 250–270.

WESTCOTT, M. R. (1968) *Towards a Contemporary Psychology of Intuition*, New York, Holt Rinehart and Winston.

WHEELER, D. (1984) 'Gatherings', *Mathematics Teaching*, **106**, pp. 24–25.

WHEWELL, W. (1858) *Novum Organon Renovatum: being the second part of The Philosophy of the Inductive Sciences*, London, John W. Parker.

WIERZBICKA, A. (1976) 'Particles and linguistic relativity', *International Review of Slavic Linguistics*, **1**(2–3), pp. 327–367.

WILDER, R. (1965) *The Foundations of Mathematics*, second edition, New York, John Wiley.

WILLIAMS, R. A. J. (1995) 'Mathematical Conversations', unpublished essay, Cambridge, Homerton College.

WILLIAMSON, T. (1994) *Vagueness*, London, Routledge.

WILLS, D. D. (1977) 'Participant deixis in English and baby talk', in Snow, C. and Ferguson, C. (eds), *Talking to Children*, Cambridge, Cambridge University Press, pp. 271–308.

WOOD, D. J., BRUNER, J. S. and ROSS, G. (1976) 'The role of tutoring in problem solving', *Journal of Child Psychology and Psychiatry*, **17**(2), pp. 89–100.

WOODS, P. (1980) 'The development of pupil strategies', in Woods, P. (ed.), *Pupil Strategies: explorations in the sociology of the school*, London, Croom Helm, pp. 11–28.

ZADEH, L. (1965) 'Fuzzy sets', *Information and Control*, **8**, pp. 338–353.

ZANDVOORT, R. W. (1965) *A Handbook of English Grammar*, third edition, London, Longman.

ZASLAVSKY, C. (1973) *Africa Counts: number and pattern in African culture*, Boston, MA, Prindle Weber and Schmidt.

Author Index

Ackermann, W. 53
Ainley, J. 131, 146–7, 158
Atkinson, J. M. 90, 92
Atkinson, P. 14
ATM 48 n.1, 50, 77
Austin, J. L. 63–4, 78–9

Bachelard, G. 30
Baker, A. 43
Balacheff, N. 35–7, 39, 106, 141, 197
Barham, J. C. 51
Bar-Hillel, Y. 70 n.12
Barrett, M. 157–8
Bell, A. W. 20
Berger, P. 157, 170 n.5
Bernays, P. 69 n.2
Bernstein, B. 11
Bernstein, R. 62
Bills, L. 40–1, 46
Bishop, A. 56
Black, M. 57
Brenner, M. 8
Brockway, D. 136
Brown, G. 1, 19 n.6, 72, 93, 183
Brown, H. J. Jnr 145
Brown, P. 58, 87, 89, 135, 137, 161, 199
Brown, R. W. 87, 109
Brown, S. 96
Buchanan-Barrow, E. 157
Burgess, R. G. 16–17
Burn, R. P. 37, 48, 29 n.5
Burns, L. C. 63, 70 n.9
Burroughs, G. E. R. 101
Burton, L. 33
Buxton, L. 116

Carlson, L. 135–6
Chafe, W. 108
Champagne, A. B. 29

Channell, J. M. 5, 63, 67, 107, 132–3, 139, 140, 142, 145, 147, 166, 184
Christie, A. 136
Clayton, J. G. 145–6, 148, 165, 167
Coates, J. 65, 139
Cobb, P. 73, 118, 148, 197, 210
Cornu, B. 30, 135
Coulthard, R. M. 81, 90
Crystal, D. 166
Cupitt, D. 13

DES xi, 50, 114 n.8
DFE 114 n.8, 145
Davenport, H. 43
Dawson, S. 54, 141, 211–12
Delderfield, R. F. 76
Deuchar, M. 73
Dewey, J. 62
Donaldson, M. 159
Drew, P. 90–2, 193
Dubois, B. L. 5
Dubois, J. 64
Durkin, K. 51

Edwards, R. P. A. 100–1, 159
Eisenhart, M. A. 14
Ellis, B. 146
Ernest, P. 54, 56

Fanshel, D. 90, 116
Farnham, D. 2
Fauvel, J. 77
Fawcett, R. P. 76, 100–1, 114 n.9, 159
Fisch, M. H. 57
Fischbein, E. 13, 29, 30
Fletcher, T. J. 48 n.1
Foxman, D. D. 1, 9, 48 n.1
Frankel, R. 90
French, P. 164
Freudenthal, H. xi, 16, 18, 52
Fuson, K. C. 148–9, 160

Gallistel, C. R. 148, 151, 160, 192
Garfinkel, H. 130, 201
Garrod, S. C. 108
Gattegno, G. 96
Gazdar, G. 18 n.1, 84, 89
Gelman, R. 148, 151, 160–1, 192
Giarelli, J. B. 18 n.1, 62
Gibbon, V. 100–1, 159
Gilman, A. 74, 87, 109
Ginsburg, H. P. 2–3, 7–9, 12, 118
Glaser, B. G. 15
Glasersfeld, E. von xii, 55–6, 72
Goffman, E. 2, 86
Goguen, J. A. 51, 58, 70 n.9
Goody, E. 86
Gordon, D. 80
Graham, D. 108
Granville, G. 34
Green, D. 27
Grice, H. P. 81–2, 84, 89
Griffiths, H. B. 21

Halliday, M. A. K. 2, 5, 64–5, 70 n.11
Hardcastle, L. 52
Hardy, G. H. 52, 65
Harel, G. 28, 29, 182
Harré, R. 77, 98, 113, 208
Hart, K. M. 152
Hempel, C. G. 32
Heritage, J. 90–1
Hersh, R. 38, 54
Hewitt, D. 23, 39, 40, 144 n.10, 180, 210
Heyting, A. 53
Hodge, R. 74, 109
Hofland, K. 114 n.5
Holland, J. H. 24, 31
Holt, J. 164
Hopper, R. 91, 93
Horn, L. 94 n.4
Howes, D. 100–1, 114 n.5, 160
Hymes, D. 5

Inhelder, B. 159

Jensen, E. M. 151
Joffe, L. 10, 165
Johansson. S. 114 n.5
Johnson, P. 56

Kane, R. B. 51
Katz, I. 34
Keats, J. 48
Kilpatrick, J. 14
Kosko, B. 57, 69 n.3, 69 n.4, 70 n.9
Kress, G. 74, 109

Laborde, C. 152
Labov, W. 90, 115–16
Lakatos, I. xii, 20, 53–4
Lakoff, G. 4, 58, 60, 67, 80, 136, 176
Lakoff, R. 116, 136, 167
Laplace, P. S. 20
Leech, G. 61, 71, 73, 206 n.5, 216
Lerman, S. 56
Levinson, S. C. 5, 58, 66, 79, 80–1, 85, 86, 87, 89, 92, 93, 94, 94 n.3, 135, 137, 144 n.9, 161, 199
Love, E. 23, 182
Luckmann, T. 157, 170 n.5
Lukasiewicz, J. 57
Lyons, J. 18 n.1, 72, 129

Mackie, A. 7
MacLure, M. 164
Maher, C. A. 144 n.8
Mason, J. 39, 41, 55, 116, 120
Mey, J. 66, 77, 79, 82, 89, 93
Mill, J. S. 31
Morgan, C. 229
Morris, C. W. 18 n.1
Moxey, L. M. 113, 208
Mühlhäusler, P. 77, 98, 113, 208

Nelson, R. B. 49 n.6
Nicholson, A. R. 52
Niven, I. 49 n.5
Nunes, T. 149, 192

Orton, A. 29, 52
Otterburn, M. K. 52

Paley, V. G. 10
Peirce, C. S. 15, 18 n.1, 30, 61–2, 175
Perera, K. 136, 182
Perkins M. R. 76, 100–1, 114 n.9, 159
Piaget, J. 7–8, 10–11, 17, 19 n.2, 56, 159
Piérrault-Le Bonniec, G. 159

Pimm, D. 6, 38, 39, 51, 68, 90, 97–8, 101, 120, 124, 143 n.1, 143 n.2, 182, 200, 208
Polya, G. 24, 26, 28, 32–5, 38–9
Prince, E. F. 5, 59, 60, 65, 67, 138–9, 147

Rees, A. 74
Rescher, N. 24
Rinsland, H. D. 100
Rolf, B. 70 n.9
Rorty, R. 62
Rosch, E. 166
Rothery, A. 51
Rowland, T. 9, 22, 40–1, 47, 51, 97, 111, 144 n.10, 195
Russell, B. A. W. 62, 70 n.9

Sacks, H. 91, 130, 201
Sadock, J. M. 67, 138, 209
Sanford, A. J. 108, 113, 208
Sankoff, G. 76
Saran, R. 15
Schegloff, E. A. 90–1
Schiffrin, D. 78, 93, 136
Schumann, H. 27
Searle, J. R. 80
Semadeni, Z. 40–1, 46
Sewell, B. 115
Sfard, A. 209, 217 n.1
Shire, B. 51
Shuard, H. 20, 51
Siegel, W. S. 146, 152
Simons, H. 74
Sinclair, J. McH. 81, 90
Singh, S. 34
Skemp, R. R. 27
Sowder, J. 146
Sperber, D. 70 n.9, 83, 94 n.4
Stamp, M. 31
Steffe, L. P. 6, 72, 148, 160, 192
Stephany, U. 65, 150, 159
Stern, G. 78

Strauss, A. L. 15
Stubbs, M. 3, 65, 67, 129, 138, 150, 169
Szeminska, A. 8

Tall, D. 28, 29, 182
Thompson, I. 148
Tiegen, K. M. 63
Tirosh, D. 29
Turquette, A. R. 57
Tuson, J. A.

Ullmann, S. 51, 63

van den Brink, J. 169 n.1

Wales, K. M. 98, 114 n.2
Wales, R. 77
Walkerdine, V. 159, 164, 170 n.7, 192–4
Wallwork, J. F. 171
Walther, G. 40, 46
Watson, A. 22, 116, 118
Watson, R. 25
Weiner, B. 147
Wells, C. G. 159
Westcott, M. R. 30
Wheeler, D. 117
Whewell, W. 23, 34
Wierzbicka, A. 136, 143 n.5
Wilder, R. 56
Williams, R. A. J. 187–9, 214
Williamson, T. 70 n.9
Wills, D. D. 98, 102
Wilson, D. 3, 70 n.9, 83–4, 95 n.4
Wood, D. J. 142
Woods, P. 158

Yule, G. 1, 19 n.6, 72, 93, 183

Zadeh, L. 57, 58
Zandvoort, R. W. 75, 109
Zaslavsky, C. 148
Zuckerman, H. S. 49 n.5

Subject Index

abduction 15, 48 n.2
absolutism 52–4
adjacency pairs 90–3, 137
affect 94, 112–13, 115–18, 201, 203
ambiguity 50, 63, 101, 200
analogy 32, 37–8, 40
anaphora
   see pronoun
approximator
   see hedge
Assessment of Performance Unit (APU) 9,
   19 n.4, 48 n.1
assimilation 27, 55
Association of Teachers of Mathematics
   (ATM) 48 n.1, 77, 117
attitude 64, 83, 133, 147
attribution
   see shield

belief 64, 112, 210
   tentative 36, 48, 54, 66

CA
   see conversation analysis
cataphora
   see pronoun
class membership 58, 60
clinical interview 6–11
   see also frame
Cockcroft Report 50, 52, 117, 182
cognitive
   conflict 104–5
   environment 84
   reference point 166
   vulnerability 210
commitment 58, 64, 66, 80, 85, 117, 160
common words
   see words
communication 50, 83
   competence 158, 216
   function 77

commutativity 104, 143 n.4
confirming instance 25–7, 40, 43, 176
conjecture 22, 26, 31, 36, 54, 60, 68, 93,
   105, 117, 120, 132, 138, 139, 141,
   173–4, 189–90, 198, 210–11, 213
conjecturing atmosphere 55, 116–17, 186,
   188, 191, 206, 215
constructivism 55–6
   theory 197
   view of learning 210
context 78, 88, 90, 102, 109
contingency 8–9, 119–20
contingent
   expression 78
   interview 118
   question 104, 131, 192
   questioning 96, 118, 131
   see also clinical interview
conversation
   expert–novice 207
   formulation 130, 199, 201–2
conversation analysis (CA) 89–93
cooperative principle (CP)
   see maxims of conversation
counting 146, 148–9, 158, 160, 169 n.1
   'how many?' 149, 162, 165, 168,
   191–2
   principles 148, 162, 192
   standard number word sequence (SNWS)
   192–3

deduction 15, 23, 48 n.3
deference 67, 88
deixis 2, 3, 77–8, 94 n.3, 102, 104–9
   conceptual 104, 113, 208
   participant 102
   procedural 183
   referent 175
   temporal 77
detachment 14, 19 n.8, 112–13, 158,
   200

discourse 78–93, 135
  analysis 89, 93
  pragmatic approach 206, 216
discovery 32, 117
dispreferred turn 92, 137, 174, 176, 179,
  203
double entendre 114 n.6
doubt 64

ellipsis 101, 206 n.6
emotion 116
  see also affect
enculturation 51, 56, 200
enthymematic 25–6, 68
  leap 176, 197
  premises 26, 35
epistemological obstacle 30
estimation 124, 145–69
Euler, Leonard 34, 49 n.5
examples 35–6
explanation 38–48, 119, 125, 136, 141
  speech 110, 141, 177, 185
extrapolativity 30, 36, 196

face 56, 86, 140, 190, 207
  negative 86, 173
  positive 86, 88, 173, 176
  wants 178, 190
face threatening act (FTA) 86, 89, 93, 173,
  177, 184, 199–201, 203
  see also politeness
fallibilism 53–5, 62
fallibilistic 54
  way of teaching 54, 141
fallible 54, 68, 212
false start 199, 203
Fawcett corpus
  see language corpora
fear of error 115
  see also affect
Fermat's Last Theorem 34, 194
Fibonacci sequence 41, 188
figures of speech 84
focus 108, 113, 208
formulation
  see conversation
frame 11, 19 n.6, 191
  weakly-framed 11, 96–7, 119, 120
  strongly-framed 11

FTA
  see face threatening act
fuzziness 56–8, 141
  see also vagueness

gambit 119, 143 n.2
Gauss, C. F. 39, 111
gender difference 167
generalization 18, 20–34, 61, 74–5, 101–3,
  109–13, 118, 120, 124, 133, 176, 186
  disjunctive 28–9
  empirical 40, 46
  expansive 28, 172–4, 177, 181, 189–90
  reconstructive 28, 182
  structural 40, 46
  see also vague
generic example 38–48, 112, 120, 123, 143
  n.4, 177, 202
Grice, H. P. 123, 125, 136

hedge 57–9, 67, 85, 115–43, 152, 180,
  187, 196, 200, 209–11, 213, 216
  adaptor 60–1, 124–5, 142, 174
  approximator 67–8, 124, 131, 138
  categories 57–9, 123, 142
  epistemic 140, 150, 173–4, 178, 209
  maxim 135–7, 197
  performative 81, 128, 138
  preferred 144 n.8
  prosodic 126
  root 139, 178
  rounder 60–1, 69 n.1, 124, 132, 142
  taxonomy 138–40
  see also maxims of conversation; shield
hedges
  'about' 127, 131–4, 140–1, 166, 184
  'approximately' 135
  'around' 131–4
  'basically' 134–5, 144 n.8, 161–3, 168
  'maybe' 125–31, 159, 161–3, 168
  'perhaps' 127, 160
  'possibly' 127
  'probably' 184
  'quite' 174
  'really' 133, 183
  'roughly' 133
  'sort of' 199
  'think' 125–31, 159–61, 163, 178
Hempel's paradox 32

hesitation 137, 143 n.5, 174, 198, 199
Howes corpus
  see language corpora

implicature 60, 81–5, 125, 139
indirectness 3, 85–6, 88, 178, 184, 191,
  195, 200
  conventional 88
induction 15, 20–34, 48 n.3, 91, 93
  conjecture 197, 210
  constraint 30–2
  default hierarchies 30–2
  inference 33, 40–1, 68
  proof by 38, 48 n.2
Informal Research Group 171, 212
information-in-hand 22, 27
inhibition 116
Initiation–Response–Feedback (I–R–F)
  cycle 90–1
institution
  see school
intention 66–8, 136
  interactional 205
  transactional 205
interaction 51, 64, 72, 206–7
interpretation 13–14, 16, 64, 66, 71, 91,
  157, 215
  truth-conditional 144 n.9
intersubjectivity 73, 91
interview 119, 153
  see also frame; transcription
intonation 10, 91, 126, 165, 168–9
  see also prosody
intuition 29–30, 48, 53, 189
intuitionism 53, 56
inverse
  mapping 108
  operation 107
investigation 23, 33, 117, 173, 181–2, 187
  'stairs' 41–3, 186
  'partitions' 21–2, 38–9
I–R–F
  see Initiation–Response–Feedback

jelly-baby effect 165–6
Jonathan 17, 198–205, 206 n.7

knowledge 19 n.7, 53–4, 56, 64, 141, 158

Lagrange's Theorem 28, 40
language 1, 5, 50–1, 56, 58, 60, 62, 67, 78
  81, 200, 207
  corpus 112, 160
  Marked 159–61, 196
  philosophy of 61–4
language corpora
  Fawcett corpus 76, 114 n.9, 136, 159
  Howes' corpus 114 n.5
  Lancaster–Oslo–Bergen (LOB) corpus
    114 n.5
  Make Ten corpus 120, 137
  Susie corpus 114 n.9
linguistic
  enculturation 194
  formula 143 n.4
  phenomena 205
  pointer 101, 108, 118, 210
  principle 2, 6
  repertoire 145
  toolkit 171

'Make Ten' 119–23, 192, 196
  see also language corpora
Marker 152–3
  primary 153, 161
  secondary 153, 156, 168
mathematical processes 20–1, 241
maxims of conversation 82, 84
  cooperative principle (CP) 82
  flouting 83–4, 130, 134
  hedge 136, 174
  Manner 82, 85, 135–6, 174
  Quality 82, 84, 95 n.5, 147, 161, 199
  Quantity 82, 84, 93, 107, 136, 139,
    144 n.9, 147, 149, 166, 174
  Relevance 82, 130
  violation 85, 136
  'well' 92, 135–6, 143 n.5, 174, 176,
    191, 197, 199–200, 205
meaning 12, 56, 63, 65–6, 83, 129, 141,
  159, 194, 197
  compatibility 72–3
measuring in school 147
modal
  auxiliary 65, 150, 168
  concept 159
  form 150

language 159
logic 53
modality 61, 64–6, 209–11
  'can' 159
  'could' 159
  'may' 161
  'might' 159, 161, 168
'more' 190, 193, 214

Number Theory 34–48, 62, 69 n.5, 110,
  112, 181, 198–200, 217 n.2

ontogenesis 53
ontological reality 55
Open University 116, 170 n.5, 195

pattern spotting 23, 40
Piaget, J. 6–7, 17
Platonism 53, 69 n.2, 103
plausibility
  see shield
plausible 24–5, 33, 68
pointer 112–13, 118, 143
politeness 65, 85–7, 140, 202, 206
  strategies 86
power 73–4, 87, 109, 157, 207–8
pragmatic 18 n.1, 52, 57–8, 68, 136
  analysis 216
  discourse 206, 216
  goal 140, 147, 206
  interpretation 216
  modality 159
  meaning 136
pragmatics 5, 15, 18 n.1, 59, 61, 66–8, 71,
  78, 138
pragmatism 18 n.1, 62, 66
preciseness paradox 63
  see also Tiegen's paradox
precision 50, 52, 62, 68, 214
prediction 22, 36, 85, 115, 119–21, 124,
  133, 173, 176, 178, 181, 192
preference organization 90–3
prime number 52, 198
PrIME project 195
probe 120, 145, 152–3, 162–3, 165,
  173
problem solving 32–4

prompt 195, 199
pronoun 73–4, 96–109, 209–9, 216
  alternation 76
  anaphora 73, 77, 94 n.3, 102, 108, 175,
    196
  cataphora 73, 101–2
  co-referential 73
  deictic role 113
  'I' 74, 187, 209
  impersonal 102, 110
  imprecision of referent 98
  'it' 99–109, 175
  'it works' 107, 113
    see also generalization
  'on' (fr.) 76
  pronominal reference 108
  rhetorical distancing 114 n.2, 200
  'thou' 74, 94 n.2
  'tu-vous' distinction 74, 76
  'we' 74, 97–8, 113, 114 n.2, 178, 181,
    187, 191, 200, 204, 208
  'you' 74–6, 109–13, 181, 183, 185, 187,
    200, 204, 209
proof 35, 38–48, 53–4, 120, 198, 202–3
  action 40–1
  crucial experiment 36–7, 106, 175
  diknumi 77
  by Mathematical Induction 38, 48 n.2
  naïve empiricism 35, 46
  'proper' 46
  without words 49 n.6
  see also generic example
propositional attitude 3, 62–3, 133, 136,
  139, 169, 181, 190, 206, 210, 213
  reasoning 132, 152, 164, 166–7
prosody 126, 166, 168, 192
  statement-tag 191, 206 n.2
  see also hedge

quasi-empirical 53–4, 141, 173
  teaching 173, 177
  see also fallibilism
question
  directing 131
  genuine question 186
  style 179
  tag-question 187
  testing 115, 131, 162

reference 72–3
 ambiguity 101
refutation 53–4
regularity 93
risk 132, 143, 143 n.6, 147, 150, 188
 taking 115, 117, 148, 167, 173
round number 133, 140, 165, 184
rules 120, 122

SAV
 see speech act verb
scaffolding 142, 188, 198
schema 27, 55
school 171
 apprehension of 157, 162, 164
 as institution 157
self-correction 136, 163, 180, 193
self-esteem 88, 116
semantics 78, 138
 truth-conditional 18 n.1, 26, 57–8, 66,
  68, 138
sequence
 arithmetic 188
 geometric 188
shield 59, 125, 132, 138, 141, 147, 173,
 191, 204
 attribution 60–1, 124, 142, 177, 196
 epistemic 184
 plausibility 60, 121, 123, 142, 161,
  187
Simon 11, 96, 110, 118–20, 134, 136
SNWS
 see standard number word sequence
socialization 158, 170 n.5, 207
solidarity 73–4
sorites paradox 61, 70 n.9
speech acts 78–88
 felicity conditions 79–80
 force 79
 performative verbs 79–81
 indirect 80, 88, 170 n.4, 172, 209–11
 sincerity condition 79, 115
 speech act verb (SAV) 79–81
standard number word sequence (SNWS)
 see counting
supervision 198, 202, 205

Susie 11, 17, 96, 99–113, 118, 120, 183
 see also language corpora
syntax 78, 138, 200

tape recorder 120, 153, 198–99
Tiegen's paradox 63, 147
transcription 10, 14, 91, 125, 171, 179–80,
 186, 214–15
 conventions 218
 see also discourse
triangular number 110–11

uncertainty 67, 116–18, 133, 152, 188, 214
undergraduate mathematics 41–7, 198–205
utterance 69 n.7, 78–9, 101, 191

vague
 estimate 196
 generaliser 109, 111, 113, 175
 language 58, 146, 195, 213
 predicates 58
 referent 109
vagueness 3, 50–68, 86, 124, 216
 definition 61
 exploitation of 114 n.6
verb
 forms 81
 modal verbs 81
 performative
  see speech acts
 private 129
vulnerability 116, 143, 147

Wiles, Andrew 34, 194
wisdom 207
words
 ambiguity 50–2
 common 101
 corpus 100
 frequently-used 100–1, 112, 114 n.4, 136
 occurrence 100
 vocabulary 101, 159

Zone of Conjectural Neutrality (ZCN)
 141–2, 144 n.10, 176, 181, 183, 188,
 210–12, 216
zone of proximal development 144 n.10

Printed and bound by CPI Group (UK) Ltd, Croydon, CR0 4YY

17/10/2024

01775685-0016